Cell culture techniques are invaluable to the modern researcher but difficult to carry out successfully. As part of the series of Handbooks in Practical Animal Cell Biology, this volume offers a concise practical guide to the basic essentials of the technique.

Researchers new to cell culture will find a clear explanation of the essential equipment of a tissue culture facility, including tissue culture media and sera. It describes methods for growing suspension and adhesion cultures, including how to store cells and prepare primary cultures from tissues. For those already culturing cells, the handbook will act as a handy reference to the basic techniques.

The essence of the book is to deal with the generalities of cell culture to give a grasp of the basic concepts before involvement in more specialized work in the field. Ideal for anyone moving into tissue cell culture techniques or looking for a concise reference book.

General techniques of cell culture

Handbooks in Practical Animal Cell Biology

Series editor:
Dr Ann Harris
Institute of Molecular Medicine, University of Oxford

Already published in the series:

Epithelial cell culture, edited by A. Harris

Endothelial cell culture, edited by R. Bicknell

Forthcoming in the series:

Marrow stromal cell culture, edited by M. Beresford and J. Owen

Haemopoietic and lymphoid cell culture,
edited by M. Dallman and J. Lamb

Endocrine cell culture, edited by S. Bidey

General techniques
of cell culture

Maureen A. Harrison
Central Cell Services, Imperial Cancer Research Fund,
Lincoln's Inn Fields, London, UK

and

Ian F. Rae
Fermenter Service, Imperial Cancer Research Fund,
Clare Hall Laboratories, Blanche Lane, South Mimms,
Hertfordshire, UK

CAMBRIDGE
UNIVERSITY PRESS

PUBLISHED BY THE PRESS SYNDICATE OF THE UNIVERSITY OF CAMBRIDGE
The Pitt Building, Trumpington Street, Cambridge CB2 1RP, United Kingdom

CAMBRIDGE UNIVERSITY PRESS
The Edinburgh Building, Cambridge CB2 2RU, United Kingdom
40 West 20th Street, New York, NY 10011-4211, USA
10 Stamford Road, Oakleigh, Melbourne 3166, Australia

First published 1997

Typeset in 10½/13 Bembo

A catalogue record for this book is available from the British Library

Library of Congress cataloguing in publication data

Harrison, Maureen A. (Maureen Anne), 1948–
    General techniques of cell culture / Maureen A. Harrison and Ian
F. Rae.
        p.   cm. – (Handbooks in practical animal cell biology)
    Includes index.
    ISBN 0 521 57364 5 (hb). – ISBN 0 521 57496 X (pb)
    1. Cell culture – Laboratory manuals.   I. Rae, Ian F. (Ian
Fraser), 1939–   . II. Title. III. Series.
    QH585.2.H37   1997
    571.6′38–dc21   97–6516   CIP

ISBN 0 521 57364 5 hardback
ISBN 0 521 57496 X paperback

Transferred to digital printing 2001

Contents

Preface

Handbooks in Practical Animal Cell Biology is a series aimed to provide practical workbooks on specific primary cell lineages. Contributing authors to each volume have assumed that their readers will have a basic level of skill and knowledge of general cell culture techniques. The role of this volume on general cell culture techniques is two fold: First, to provide a grounding in the basics of cell culture techniques and second, to give practical details of setting up a cell culture facility from scratch or improving an existing one. The authors run a large and efficient cell culture facility at the Imperial Cancer Research Fund, London; however, the procedures and equipment that they describe are equally applicable to the laboratory with one laminar flow hood and one cell culture incubator. Cell culture novices would be well advised to gain competence in the culture of a few robust long-term cell lines prior to attempting the culture of one of the more demanding primary cell types described in the individual cell lineage volumes. Those with ample experience of cell culture techniques will find this volume most valuable for troubleshooting and for ideas on how to improve their cell culture facility.

Ann Harris
Oxford October 1996

Acknowledgements

The authors gratefully acknowledge the help and advice of: Dr Ann Harris, Paediatric Molecular Genetics, University of Oxford; Mr William House, former Deputy Director of Clinical Research and former Director of Support Services, ICRF; Dr Helen Hurst, Head of Gene Transcription Laboratory, ICRF Oncology Group; Mr Stanley Fisher, Health and Safety Manager, ICRF; Mr Leslie Fullerton, Safety Officer, Clare Hall Laboratories, ICRF; Mr William Bessant, Head of Photographic Department, ICRF; Mr Stephen Wilson, Central Cell Services, ICRF; Ms Carol Kolar, Eppley Institute, Omaha, Nebraska, USA; Dr Terry Lawson, Eppley Institute, Omaha, Nebraska, USA; Sales departments and representatives of various suppliers of equipment and chemicals. Dr Ali Mobasheri kindly supplied cover photographs for the series.

1

Introduction

The aim of this volume is to provide a guide to the basic essentials of tissue culture, progressing from the equipment needed, through useful media and sera to the handling of different types of cells, their growth and storage, how to recognize and deal with contamination, and to provide some pointers towards good quality control and safe handling procedures. Detailed protocols on specialized applications are outside the scope of this book, although a chapter is included which covers the basics of some of the more widely used special techniques.

Individual volumes in this series deal with the specific cultures of primary cell types, but before commencing on a project necessitating more specialized skills it may be useful to experience the pleasures and pitfalls of basic tissue culture methods, which will be applicable to primary systems, through handling cell lines. Cell banks and smaller cell production facilities exist largely for the purpose of issuing cell lines and during the culturing of a few selected varieties, anyone of average dexterity and with a good grasp of sterile technique will soon develop a basic expertise.

Tissue culture developed from some of the embryology techniques used in the last century, which involved maintaining the medullary plate of a chick embryo in warm saline. Attempts were also made to maintain pieces of human skin *in vitro*. This was followed by attempts to maintain leukocytes from the salamander in hanging droplets (Jolly, 1903). From this early work, the traditional tissue culture techniques were rapidly devised. The term 'tissue culture' includes both cell and organ culture, although within the confines of this book we will be discussing only the former.

Harrison, in 1907, worked with frog tissues, since they require no incubation and since tissue regeneration is more common in lower vertebrates, there was a possibility that growth was more likely to occur than with mammalian tissue. Burrows's and Carrel's work (Burrows, 1910; Carrrel and

Burrows, 1912) showed that animal cells could be grown *in vitro*. A history of somatic cell genetics will provide a more detailed account of the origins of the subject (Harris, 1995).

The rudiments of tissue culture as we know it today came predominantly from Dr Earle's group at the National Cancer Institute among others. This group grew cells directly on glass, propagated cultures from single cells, and were successful in growing suspension cultures. Numerous workers followed this with investigations into the factors in culture medium that are necessary for growth. Some of these researchers have given their name to media in common use today, e.g. Fischer, Waymouth, Eagle. The development of tissue culture as a useful tool in the modern laboratory arose largely because of the needs of (a) the production of antiviral vaccines and (b) cancer research. Now that there are standard procedures for the production of large numbers of cells, and sera and media may be obtained commercially, tissue culture lends itself to a whole range of investigative fields including all aspects of cell biology, physiology and biochemistry.

The advantages of tissue culture include the ability to control the environment of the cell (pH, temperature, osmotic pressure, O_2 and CO_2). Additionally, after one or two passages cultured cell lines assume a uniform constitution, and as the cells are randomly mixed at each passage, there is a tendency for the culture conditions to produce a similar type of cell in as much that the most vigorously growing cells will predominate. The other noteworthy advantage is that, in numerous instances, tissue culture may be used to replace experimentation on live animals, which aside from ethical preferences is cheaper and quicker.

Cell markers

Characterization of cell lines is important. It is naturally a necessity to be able to correlate a culture with its original tissue from the point of view of its descent and its stage of advancement during the course of the development of the mature cell type. During the maintenance of cell lines there may be instability in the phenotype and genotype caused by variations in culture conditions, selective overgrowth or genetic changes. It is therefore important that conditions are standardized and a seed stock is preserved to be able to return to at intervals to maintain consistency. It is essential that cells are checked regularly for cross contamination (see Chapter 8). Stable markers are needed for characterizations, and culture conditions may need to be modified so that these markers are expressed.

For details of markers of other cell lines the reader should refer to other books in the Handbooks in Practical Animal Cell Biology series.

In general, the most obvious technique used to identify cells is morphology. However, it should be remembered that cellular morphology can appear different according to the culture conditions, e.g. Mouse 3T3 cells have a fibroblastic appearance at low cell density but when confluent become epithelial-like. Over the years the terms 'fibroblastic' and 'epithelial' have come to apply to the appearance of the cells rather than their origin. If this criterion is used, a cell which has length more than twice its width could be called fibroblastic, whilst a monolayer cell with more regular dimensions could be said to be epithelial.

Growth cycles, which give an idea of the population doubling time and the time it takes for a monolayer to become confluent or a suspension culture to reach saturation density should remain constant so that a change in the expected times, assuming that culture conditions have remained stable, could indicate cross contamination, or senescence (see Chapter 5). However, such changes may also indicate transformation (possibly due to viral infection) or contamination.

Types of cells

Cells may be loosely divided into two main types, those that grow as a suspension and those which, as the name suggests, grow as an adherent cell culture. Primary cultures and the evolution of cell lines will be discussed in Chapters 5 and 6.

Propagation of cells becomes possible due to cell proliferation. When cells are selected from a culture, the subculture is termed a 'cell strain' and usually strains will be selected for a particularly detailed characterization.

Suspension cultures are derived from cells which can divide and survive without being attached to a substrate, e.g. cells of haemopoietic lineage, whereas adherent cultures or monolayers must adhere to a substrate to survive.

It is possible to culture many different cell types of various lineages, e.g. epithelial cells including keratinocytes, mammary gland duct cells, cervical cells, cells from the gastrointestinal tract, liver, pancreas, kidney, bronchial and tracheal cells; mesenchymal cells which include connective, adipose and muscle tissue, cartilage and bone. Neuroectodermal cells include glia and endocrine cells, and the haemopoietic system includes macrophages and lymphoid cells. Specialized techniques for the culture of these cell types will be found in other books in this series.

Sources of cells

Aside from preparing cultures from the original tissues (see Chapter 6), the most usual sources of cells are cell banks or cell stocks from other workers in the field.

A number of reliable organizations produce good-quality controlled cell cultures, which have been subjected to tests for viability, the absence of contamination including mycoplasma, karyotypic and isoenzyme analysis, DNA fingerprinting, and in some cases tests for viral susceptibility, tumourgenicity, biochemical traits and drug susceptibility. Cell banks will also accept deposits of useful cell lines. It may be useful, at this point, to briefly describe the scheme for banking cells, as a similar procedure scaled down may be adopted for banking cells in a laboratory/department.

When a new cell line is brought into the laboratory, there should be defined procedures on how to handle it. Cultures should be handled in Class II safety cabinets in a quarantine laboratory away from the main tissue culture area. A small amount should be frozen as soon as possible and quality control tests carried out. When the quarantine conditions (see Chapter 8) have been satisfied, the new cultures may be transferred to the main laboratory for production of master and working banks.

An outline is given of the general plan for running a cell bank in Figure 1.1.

If an organization is attempting to run a cell bank on a smaller scale, i.e. producing cultures for that particular organization only, it will usually not be possible to fulfil the criteria as laid out in the plan above and amendments must be made to take into account such things as space available for the storage of frozen cells, the practicalities of returning samples to the originator for verification and the pressures of having to produce cultures in the shortest possible time. However, it would be foolhardy in the extreme to cut out any contamination checks (particularly for mycoplasma) or species verification.

The cell banks listed at the end of this chapter produce catalogues of material held, and access to up-to-date information is available using on-line data systems. (The American Type Culture Collection (ATCC) database is available on-line via the Cambridge, UK based MSDN computer network using electronic mail systems.) Both the ATCC and ECACC catalogues are accessible on the world-wide web:

ATCC http://www.Atcc.org/

ECACC http://www.biotech.ist.unige.it/cldb/descat5.html

Both these cell banks have begun a project to generate a European directory of cell line resources available in European laboratories.

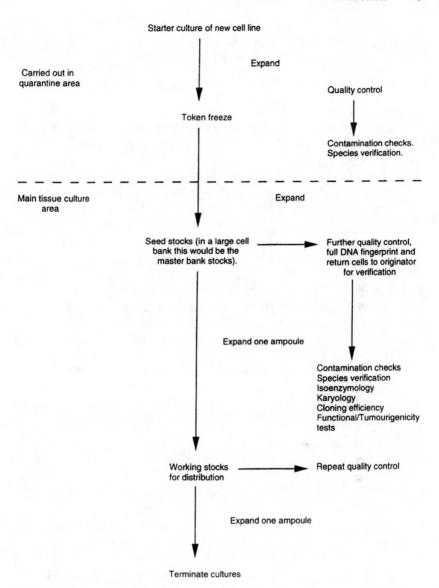

Fig. 1.1. Procedures for handling a new cell line.

Addresses of useful cell banks:

American Type Culture Collection, 12301 Parklawn Drive, Rockville, MD 20852, USA   Tel: (301) 991 2600.   Telex: 908768 ATCC ROVE.   Fax: (301) 231 5826

The European Collection of Animal Cell Cultures, PHLS Centre for Applied Microbiology and Research, Porton Down, Salisbury, Wiltshire SP4 OJG, UK   Tel: 01 980 610391.   Telex: 47683 PHCAMR G.   Fax: 01 980 611315 E-mail: Telecom Gold, 75:DB1 0222

NIGMS Human Genetic Mutant Cell Repository, Coriell Institute for Medical Research, 401 Haddon Avenue, Camden, NJ 08103, USA   Tel: (609) 757 4848.   Fax: (609) 964 0254

References

Burrows, M. T. (1910). The cultivation of tissues of the chick-embryo outside the body. *J. Am. Med. Assoc.*, **55**, 2057–8.

Carrel, A. & Burrows, M. T. (1912). Cultivation of adult tissue and organs outside the body. *J. Am. Med. Assoc.*, **55**, 1379–81.

Harris, H. (1995). *The Cells of the Body: A History of Somatic Cell Genetics.* Cold Spring Harbor Press, USA.

Harrison, R. G. (1907). Observations on the living developing nerve fibre. *Proc. Soc. Exp. Biol. (NY)*, **4**, 140.

Jolly, J. (1903). Sur la durée de la vie et de la multiplication des cellules animals en dehors de l'organisme. *CR Soc. Biol. (Paris)*, **55**, 1266.

2

Equipment

In its broadest definition, equipment includes the laboratory in which cell culture work is undertaken. Some are fortunate enough to occupy purpose-built cell culture facilities but many use existing laboratories which require varying degrees of adaption to house a successful cell culture area. When planning a laboratory for cell culture, six main functions have to be accommodated. These can be neatly divided into two main groups: sterile handling and support services. Sterile handling includes a cell culture and manipulation area which should be adjacent to an incubation and a storage area. Support services include washing-up, preparation (repackaging) and sterilization. These three functions should also be adjacent to each other and provision made to extract the large amounts of heat and steam associated with this type of operation. It is not essential for the services to be adjacent to the sterile handling area but they should be within the same building. By far the most important consideration is to minimize the chances of microbiological contamination of cell cultures. One of the main causes of contamination can be sudden draughts of room air crossing the work surface from opening doors, the passage of staff behind the operator, open windows or wall-mounted air-conditioning units. Where a laboratory has opening windows, it is vital they are kept closed whenever cell culture work is in operation. Wall-mounted air-conditioning units have no place in a cell culture laboratory because the damp internal conditions harbour and support a source of microbiological contamination readily circulated by the forced movement of air from the unit. It is very important therefore, that the area designated for handling and manipulating cells should be as far away as practicable from the laboratory entrance and from the main passage of staff within the laboratory. Much of the other equipment required for cell culture work will be common to most laboratories, e.g. various types of glassware (pipettes, beakers, flasks, bottles, measuring cylinders, etc.) all of which need to be sterile; water baths, centrifuges, balances, etc. The number of items of specialized equipment required are comparatively few.

Large equipment

Laminar flow cabinet

Although it is possible to culture cells on an open bench employing careful aseptic technique with or without the aid of a gas burner, the use of a laminar flow cabinet significantly reduces the chance of contamination and eliminates the need for a gas burner. If using a laminar flow cabinet, the minimum requirement is for a horizontal laminar flow cabinet primarily designed to protect the work from the operator but not the operator from the work. With ever-increasing improvements in health and safety regulations in some countries this type of cabinet is now considered to be insufficiently protective of the operator and therefore cannot be recommended. Much better protection for both the operator and the work is provided by a Class II vertical laminar flow (Fig. 2.1) cabinet and this type of cabinet is used widely in cell culture laboratories. Two main variants are available, both employing high efficiency particulate air (HEPA) filters with an efficiency of 99.999%: (a) Class II microbiological safety cabinet (MSC) designed to BS5726 (1992) (UK), NSF 49 (USA), DIN 12950(Germany), NFX 44-201 (France) or AS2252 (Australia) suitable for working with Category 2 pathogens. These are part-open fronted cabinets with double HEPA filters that give sufficient protection to permit them to be used without external ducting. Air is drawn in through a grill at the front lower edge of the cabinet and circulates through a HEPA filter before passing vertically down across the internal front face of the cabinet as a curtain of air at 0.4m/s thereby protecting both the operator and the work. In this type of cabinet, 70% of the air is recirculated through one HEPA filter whilst the remaining 30% is discharged through a second HEPA filter to the external atmosphere. (b) Tissue culture cabinet specifically designed for cell culture and other low risk operations but not suitable for work with pathogens. These are less expensive than the Class II MSC but still meet the safety regulation requirements.

All cabinets should be performance tested at least once a year by a competent person who will advise on filter changing. Some laboratories consider it preferable to leave cabinets running all the time, but, where they are not, the cabinet should be isolated by closure of the operator access, i.e. by replacing the removable front panel.

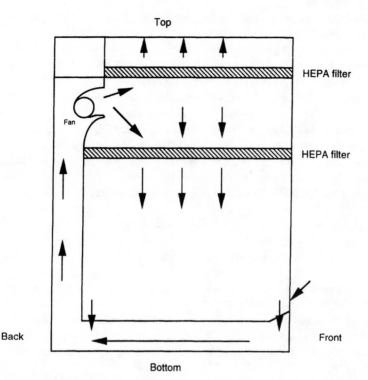

Fig. 2.1. Side view of a typical laminar flow cabinet design in which 70% of the air is recirculated. Room air is drawn in at the bottom front edge of the cabinet, mixed with air being drawn from the working area and circulated round the rear ducting to the upper chamber. Here, 30% of the air is returned to the room through one set of HEPA filters and the remaining 70% recirculated into the working chamber through a second set of HEPA filters,

Manufacturers of laminar flow cabinets

Astec Environmental Systems, 30–31 Lynx Crescent, Weston-Super-Mare, Avon BS24 9BP, UK   Tel: 01934 418685   Fax: 01934 419033

The Baker Co. Inc., PO Drawer E, Sanford, Maine 04073, USA.   Tel: (207)-324-8773   Fax: (207)-324-3869

Bassaire Ltd, Duncan Road, Swanwick, Southampton, Hampshire SO3 7ZS, UK   Tel: 01489 885111   Fax: 01489 885211

Bigneat Ltd, 5 Pipers Wood, Brambles Farm Industrial Estate, Waterlooville, Hampshire PO7 7XU, UK   Tel: 01705 266400   Fax: 01705 263373

Coy Laboratory Products Inc., 14500 Coy Drive, Grass Lake, Michigan 49240, USA   Tel: (313)-475-2200   Fax: (313)-475-1846

Envair Ltd, York Avenue, Haslingden, Rossendale, Lancashire BB4 4HX, UK   Tel: 01706 228416   Fax: 01706 831957

Gelaire – supplied by ICN Ltd.

Heraeus Instruments GmbH, Heraeusstrasse 12–14, PO Box 15 53, D-6450 Hanau 1, Germany   Tel: (06181)35-465   Fax: (06181)35-749

Heto-Holten Laminair A/S, Gydevang 17–19, DK-3450 Allerod, Denmark.   Tel: (+45) 48 14 27 77   Fax: (+45).42.27.46.55

ICN Biomedicals Ltd, Unit 18, Thame Business Centre, Wenman Road, Thame, Oxfordshire OX9 3XA, UK   Tel: 01844 213366   Fax: 01844 213399

ICN Pharmaceuticals Inc., 3300 Hyland Avenue, Costa Mesa, California 92626, USA   Tel: (714)-545-0113   Fax: (800)-334-6999

Jouan Inc., 110B Industrial Drive, Winchester, Virginia 22602, USA   Tel: (540)-869-8623   Fax: (540)-869-8626

Jouan Ltd, Merlin Way, Quarry Hill Road, Ilkeston, Derbyshire DE7 4RA, UK   Tel: 0115 944 7989   Fax: 0115 944 7080   E-mail: jouan.co.uk

Jouan SA, CP.3203 – 44805 Saint-Herblain Cedex, France   Tel: 40.16.80.00   Fax: 40.94.70.16

Labcaire Systems Ltd, 15 Hither Green, Clevedon, Avon BS21 6XU, UK   Tel: 01275 340033   Fax: 01275 341313

Medical Air Technology (MAT), Wilton Street, Denton, Manchester, M34 3LZ, UK   Tel: 0161-320 5652   Fax: 0161-335 0313

Microflow Dent and Hellyer (MDH), Walworth Road, Andover, Hampshire SP10 5AA, U.K   Tel: 01264 36211   Fax: 01264 356452

Nu-Aire Inc., 2100 Fernbrook Lane, Plymouth, Minnesota 55447, USA   Tel: (612)-553-1270   Fax: (612)-553-0459

Incubators

Culturing cells from any organism requires the provision of environmental conditions that mimic as closely as possible those experienced by cells *in vivo*. For mammalian cells, this includes incubation at a temperature as close as possible to the normal temperature of the animal species from which the cells were taken. Generally, this ranges from 33 °C to 43 °C depending on the ambient temperature. In practice, most incubators used for cell culture are set to 35 °C to 37 °C which, for most types of mammalian cell, is satisfactory. Incubators range from what is essentially a simple insulated metal box with a door and basic temperature controls to water-jacketed CO_2 incubators with sophisticated electronic controls. The CO_2 incubator, which is designed specifically for cell culture in open vessels, e.g. Petri dishes, microtitre plates, etc. is provided with a supply of CO_2 either from a piped system or from gas cylinders. The electronic controls on the incubator ensure the atmosphere within the incubator chamber is maintained at a constant enhanced 5% or 10% CO_2. Because of the open vessels, the chamber has to be humidified to minimize fluid loss due to evaporation. This, in itself, presents an increased risk of contamination, particularly from moulds, and some are lined with copper or have sterilizable liners to help minimize the risk. For the majority of purposes a CO_2 incubator is not essential if closed vessels, e.g. flasks or roller bottles, are used. These can be flushed out with a gas mixture containing either 5% or 10% CO_2 in 95%/90% air and once sealed have their own micro-climate in which the cells can grow. In either case, the need for CO_2 can be eliminated by the use of the zwitterionic buffer HEPES that enables cells to be cultivated even in open vessels without the use of an atmosphere containing supplemented CO_2. It is important that the incubator used has a reliable and accurate temperature controller that will maintain the temperature of the chamber within very close limits and has an upper temperature cut out fitted with an alarm.

Manufacturers of incubators

Barnstead-Thermolyne Corporation, 2555 Kerper Boulevard, Dubuque, Iowa 52004-0797, USA   Tel: (319)-556-2241   Fax: (319)-556-0695

Carbolite, Aston Lane, Hope, Sheffield, South Yorkshire S30 2RR, UK   Tel: 01433 620011   Fax: 01433 621198   'Peak' range

Carbolite Inc., PO Box 7, 110 South Second Street, Watertown, Wisconsin 53094, USA   Tel: (414)-262-0240   Fax: (414)-262-0255

Elplas Ltd, 7 Norman Way, Over Industrial Park, Over, Cambridgeshire CB4 5QE, UK   Tel: 01954 31543   Fax: 01954 31983   Enkab portable benchtop mini-incubator

Forma Scientific Inc., Millcreek Road, PO Box 649, Marietta, Ohio 45750, USA   Tel: (614)-373-4763   Fax: (614)-373-6770

Genlab Ltd, Tanhouse Lane, Widnes, Cheshire WA8 0SR, UK   Tel: 0151-424 5001   Fax: 0151-495 2197

Heraeus Instruments GmbH, PO Box 15 63, D-63405 Hanau, Germany   Tel: (+49).61.81.35.300   Fax: (+49) 61 81 35 5973

Heraeus Equipment Ltd, 9 Wates Way, Brentwood, Essex CM15 9TB, UK   Tel: 01277 231511   Fax: 01277 261856

Heraeus Instruments Inc., 111A Corporate Boulevard, South Plainfield, New Jersey 07080, USA   Tel: (908)-754-0100   Fax: (908)-754-9494

Integra Biosciences AG, PO Box 74, CH-8304, Wallisellen, Switzerland   Tel: 830.2277   Fax: 830.7852

Integra Biosciences Ltd, The Annexe, New Barnes Mill, Cottonmill Lane, St Albans, Hertfordshire AL1 2HB, UK   Tel: 01727 848825   Fax: 01727 811769   'Biocenter'

Integra Biosciences Inc., 10–H Roessler Road, Woburn, Massachusetts 01801, USA
Tel: (617)-937-8300   Fax: (617)-937-8301

Lab Impex Research Ltd, Unit 5, Kingsway Business Park, Oldfield Road, Hampton, Middlesex TW12 2HD, UK   Tel: 0181-296 1122   Fax: 0181-296 1123 'Discovery' range

Lab-Line Instruments Inc., 15th. and Bloomingdales Avenues, Melrose Park, Illinois 60160, USA   Tel: (708)-450–2600   Fax: (708)-450–0943

Laboratory Thermal Equipment (LTE.) Scientific Ltd, Greenbridge Lane, Oldham, Lancashire OL3 7EN, UK   Tel: 01457 876221   Fax: 01457 870131

LEEC Ltd, Private Road No.7, Colwick Industrial Estate, Nottingham NG4 2AJ, UK   Tel: 01602 616222   Fax: 01602 616680

LMS., The Modern Forge, Riverhead, Sevenoaks, Kent TN13 2EL, UK   Tel: 01732 451866   Fax: 01732 450127

Planer Biomed, Windmill Road, Sunbury, Middlesex TW16 5BR, UK   Tel: 01932 779997   Fax: 01932 781151

Precision Scientific Inc./Napco, 3737 West Cortland Street, Chicago, Illinois 60647-4793, USA   Tel: (312)-278-8668   Fax: (312)-227-4767

Revco Scientific, One Revco Drive, Asheville, North Carolina 28804, USA   Tel: (704)-658-2711   Fax: (704)-645-3368

RS Biotech, Tower Works, Well Street, Finedon, Northamptonshire NN9 5JP, UK   Tel: 01993 680133   Fax: 01993 680155 'Galaxy' range

Sheldon International Ltd. (Shel-Lab), PO Box 19051, Portland, Oregon 97280, USA   Tel: (503)-640–3000   Fax: (503)-640–1366

Tuttnauer, 33 Comac Loop, Equipark, Ronkonkoma, New York 11779, USA   Tel: (516)-737-4850   Fax: (516)-737-0720

Tuttnauer, POB 7191, 4800 GD Breda, Netherlands   Tel: 31 76 423 510   Fax: 31.76.423.540

Cell handling equipment

Inverted microscope

Microscopic examination of adherent cells growing as a monolayer on the surface of a culture vessel is an essential element of successful cell culture. Although it is possible to see the cells with a conventional microscope, it is much easier to examine them using an inverted microscope equipped for both bright-field and phase-contrast techniques where the objectives are mounted on a moveable head situated below the stage and the lamp is attached to a column above the stage. Most have the facility for examining both flat culture vessels and roller bottles whilst the cells remain covered with medium. Suspension cells can also be examined and counted using a counting chamber (haemocytometer). The most useful objectives for examining cell cultures are conventional/phase-contrast achromatic 10 × 0.25 and 20 × 0.35 magnification, whilst a bright-field 5 × 0.12 objective is also useful but not essential. The eyepieces should be of 10 × magnification.

Manufacturers of inverted microscopes

Helmut Hund GmbH, Postfach 21 01 63, D-35550 Wetzlar, Germany   Tel: 064 41 20040   Fax: 064 41 200444   'Wilovert' range

Leica Inc., 111 Deer Lake Road, Deerfield, Illinois 60015, USA   Tel: Toll Free 800–248-0123   Fax: (708)-405-0030

Leica Instruments GmbH, Postfach 1120, Heidelberger Strasse 17–19, D-6907 Nussloch, Germany   Tel: 49 62 24 14 30   Fax: 49 62 24 10 015

Leica UK Ltd, Davy Avenue, Knowhill, Milton Keynes MK5 8LB, UK   Tel: 01908 666663   Fax: 01908 609992

Nikon Corporation, Fuji Building, 2–3 Marunouchi 3-chome, Chiyoda-Ku, Tokyo 100, Japan   Tel: 81-3-3216-1039   Fax: 81-3-3201-5856

Nikon UK Ltd, Nikon House, 380 Richmond Road, Kingston-upon-Thames, Surrey KT2 5PR, UK   Tel: 0181-541 4440   Fax: 0181-541 4584 'Diaphot TMD & TMS'

Olympus America Inc., Precision Instrument Division, Two Corporate Center Drive, Melville, New York 30622, USA   Tel: Toll Free 800–446-5967   Fax: (516)-844-5109

Olympus Optical Company Ltd, 2-43-2 Hatagaya Shibuya-ku, Tokyo, Japan   Tel: 010 81 33 209 4821   Fax: 010 81 33 340 2201

Olympus Optical Company (UK) Ltd, 2–8 Honduras Street, London EC1Y 0TX, UK   Tel: 0171-253 2772   Fax: 0171-251 6330   IX50 and IX70 range

Pyser-SGI Ltd, Fircroft Way, Edenbridge, Kent. TN8 6HA, UK   Tel: 01732 864111   Fax: 01732 865544   'Swift' range

Carl Zeiss Germany, PO Box 1380, Carl Zeiss Strasse, D-7082 Oberkochen, Germany.   Tel: 07.364.200   Fax: 07.364.6808

Carl Zeiss Inc., One Zeiss Drive, Thornwood, New York 10594, USA   Tel: (914)-747-1800   Fax: (914)-682-8296

Carl Zeiss Jena, Division of Microscopy, Tatszendpromenade 1, D-6900 Jena, Germany   Tel: 3641.5880   Fax: 3641 5882079

Carl Zeiss Ltd, PO Box 78, Woodfield Road, Welwyn Garden City, Hertfordshire AL7 1LU, UK   Tel: 01707 331144   Fax: 01707 373210

Counting chamber (haemocytometer)

This is the simplest and most commonly used method for counting cells and assessing their viability. Originally developed to enumerate blood cells, there are several types available etched with different patterns according to whether they were designed to examine either red blood cells, white blood cells or both types. White blood cells (leukocytes) are much larger than red cells (erythrocytes), and a chamber designed for counting white blood cells is required to examine and count tissue cells in culture. Of the types generally available (Burker, Fuchs-Rosenthal, Neubauer, Improved Neubauer and Thoma), the improved Neubauer is most commonly employed (see Chapter 5). It consists of a heavy glass slide with four deep channels running parallel to each other from top to bottom forming a centre platform that is slightly lower (0.1 mm) than the outer sections of the chamber. A cover glass is placed in position over the centre platform with its edges resting on the outer sections thus forming a bridge over the central platform. Always use the toughened cover slip designed for use with a counting chamber. The cover slip needs to be placed in position using firm pressure along the two edges resting on the outer sections of the chamber until it remains in position without moving, and the rainbow effect of the Newton's rings are visible along the two edges. Only then is it the correct height above the central platform. Never put pressure on to the centre of the coverslip as this is the most common cause of breakage. Into the gap is run slowly and carefully, but without hesitation, a sample of the cells in culture diluted in a vital dye, e.g. Trypan Blue, Nigrosine using a Pasteur or microlitre pipette. The use of a counting chamber is the easiest and most useful way of examining the health and viability of all types of cells. Counting chambers should be cleaned after every use with either methanol or ethanol and either stored dry or under solvent. Disposable counting chambers are also available.

Manufacturers of cell counting chambers

Weber Scientific International, 40 Udney Park Road, Teddington, Middlesex TW11 9BG, UK   Tel: 0181-977 6330   Fax: 0181-943 4224

Manufacturers of disposable cell counting chambers

Cellvision, PO Box 1232, 1700BE Heerhugowaard, The Netherlands   Tel: +31 72 57 23657   Fax: +31 72 57 21184

Electronic cell enumerators

Over the years, several different electronically controlled automatic methods have been developed for counting cell numbers in suspension, but today the system devised by Coulter Electronics is the one most widely used. This type of machine has come a long way in the last few years with greatly improved accuracy but these machines are comparatively expensive. The principle on which they work is fairly simple: cells are drawn through a fine orifice and, as they do so, there is a change in the current flow producing a series of electrical pulses proportional to the volume of the cell. These signals are amplified and then sorted and counted. They are easy to use when the cell population is of a fairly uniform size but become much more complicated where the cell population has a wider variation in size, tends to clump in suspension or is difficult to dissociate from a monolayer into single cells.

Manufacturers of electronic cell enumerators

Coulter Electronics Ltd, Northwell Drive, Luton, Bedfordshire LU3 3RH, UK
Tel: 01582 567000   Fax: 01582 490390

Disposable plasticware

A broad range of disposable plastic vessels, specifically designed and manufactured for the cultivation of both adherent and suspension cells, is available from several manufacturers. The range of vessels includes screw-capped 'flasks' (rectangular bottles with two flat surfaces for culturing cells) usually available in four sizes (25 cm^2 (70 ml), 75 cm^2 (275 ml), 160–175 cm^2 (600 ml) and 225 cm^2 (1022 ml) growth area per side); roller bottles (with a cylindrical profile) available in three standard sizes (490 cm^2, 850 cm^2 and 1750 cm^2) and with extended ('pleated', 'ribbed' or 'serrated') surfaces giving up to 3500 cm^2 of growth area for adherent cells in the same sized bottle as the 1750 cm^2; test tubes, Petri dishes, microtitre plates, cryotubes, etc. It is very important always to use plasticware manufactured for cell culture work because the surface of the plastic has been given a moderate negative electrical charge (2×10^{14}–10×10^{14} charge/cm^2) to enable the positively charged cells to adhere to the plastic surface. Plasticware designed for bacteriological work is positively charged and should not be used.

Manufacturers of disposable plasticware for cell culture

Becton Dickinson Canada Inc., 2464 South Sheridan Way, Mississauga, Ontario L5J 2M8, Canada   Tel: (905)-822-4820   Fax: (905)-822-2644
Becton Dickinson Labware, Two Oak Park, Bedford, Massachusetts 01730, USA
Tel: (617)-275-0004   Fax: (617)-275-0043
Becton Dickinson Europe, 5 Chemin des Sources, 38241 Meylan Cedex, France.
Tel: +33.76.41.64.64   Fax: +33 76 90 19 65   'Falcon' brand

Becton Dickinson (UK) Ltd, Between Towns Road, Cowley, Oxford OX4 3LY, UK
Tel: 01865 777722   Fax: 01865 717313

Bibby Sterilin Ltd, Tilling Drive, Stone, Staffordshire ST15 0SA, UK   Tel: 01785
812121   Fax: 01785 813748

Corning brand range: obtainable from Bibby Sterilin Ltd in UK

Corning Costar Corporation, One Alewife Center, Cambridge, Massachusetts
01240, USA   Tel: (617)-868-6200   Fax: (617)-868-2076

Costar Europe Ltd, PO Box 94, 1170 AB Badhoevedorp, The Netherlands   Tel:
020 659 6051   Fax: 020 659 7673

Costar UK Ltd, 10 The Valley Centre, Gordon Road, High Wycombe,
Buckinghamshire HP13 6EQ, UK   Tel: 01494 471207   Fax: 01494 464891

Greiner GmbH, Maybachstrasse 2, D-72636 Frickenhausen, Germany   Tel: 07 022
50 10   Fax: 07 022 50 15 14

Greiner Labortechnik Ltd, Station Road, Cam, Dursley, Gloucester GL11 5NS, UK
Tel: 01453 548833   Fax: 01453 544070

Nalge Nunc International, PO Box 20365, 75 Panorama Creek Drive, Rochester,
New York 14602, USA   Tel: (716)-586-8800   Fax: (716)-586-8431

Nippon Becton Dickinson Company Ltd, Shimato Building, 5-34, Akasaka-8-
chome, Minato-ku, Tokyo 107, Japan   Tel: (81)-3 3403-9991   Fax: (81)-3
3403-5321

Nunc A/S, Kamstrupvej 90, PO Box 280, DK-4000 Roskilde, Denmark   Tel:
(+45) 46 35 90 65   Fax: (+45) 46 35 01 05   E-mail: infociety@nunc.dk

Nunc A/S, Life Technologies Ltd, 3 Fountain Drive, Inchinnan Business Park,
Paisley PA4 9RF, Scotland, UK   Tel; 0141-814 6100   Fax: 0141-887 1167

Nunc Inc., 2000 North Aurora Road, Naperville, Illinois 60563-1796, USA   Tel:
(708)-983-5700   Fax: (708)-416-2556

Sterilin brand range: see Bibby Sterilin Ltd

Glass culture vessels

Where budget restraints prevent the purchase of disposable plastic culture
vessels, glass bottles can be used instead. A range of relatively cheap flat-sided
clear soda glass prescription bottles are available: 60 ml (2 oz/32 cm^2), 100 ml
(4 oz/48 cm^2), 150 ml (6 oz /69 cm^2), 200 ml (8 oz/75 cm^2) and 500 ml (20
oz/132 cm^2) for which either aluminium caps with rubber liners (wads) or
plastic caps are obtainable. Also available are round profile 2000 ml (80 oz/690
cm^2) prescription bottles in amber glass, which can be used as roller bottles.
All these can be used either for reagent and media storage or for culturing cells.
Borosilicate glass bottles, e.g. Pyrex, Schott, are also obtainable but these are
more expensive. Other types of specialized glassware manufactured in borosilicate glass specifically for culturing cells are also available including roller
bottles and Roux culture flasks (230 cm^2). All new glassware must be thor-

oughly washed before sterilization for use to remove any soiling accrued during manufacture, packaging or transportation. Check the neck rim of every bottle for chipping before use: chipped rims may prevent the formation of a gas-tight seal between the cap and the rim and can also be the cause of medium leakage and contamination.

Manufacturers of soda glass bottles

Beatson Clark, The Glassworks, Greasborough Road, Rotherham, South Yorkshire S60 1TZ, UK   Tel: 01709 828141   Fax: 01709 828478

United Glass Ltd, Glasshouse Loan, Alloa FK10 1PD, Scotland, UK   Tel: and Fax: 01259 218822

Manufacturers of borosilicate glass bottles

Bellco Glass Inc., 340 Erudo Road, Vineland, New Jersey 08360, USA   Tel: (609)-691-1075   Fax: (609)-691-3247

'Pyrex' Brand. Obtainable from J.Bibby Science Products Ltd, Tilling Drive, Stone, Staffordshire ST15 0SA, UK   Tel: 01785 812121   Fax: 01785 813748

Schott Glaswerke, Postfach 2480, D-55014 Mainz, Germany   Tel: 06131 66 43 90 Fax: 06131 66 40 16

Wheaton Ltd, Division of Wheaton Industries, 1301 North Tenth Street, Milville, New Jersey 08332, USA   Tel: (609)-825-1100   Fax: (609)-825-1368

Carbon dioxide

Most cells are cultured using the bicarbonate/carbon dioxide buffering system, which requires the culture vessels to be flushed out with air supplemented with 5% or 10% carbon dioxide gas, prior to inoculation, depending on the amount of sodium bicarbonate present in the medium. Gas should always be delivered to a culture vessel through an appropriate reducing valve, with an in-line hydrophobic airline filter and a sterile outlet renewed prior to every occasion. Purpose-built laboratories will usually have a central pipe system but, where this is not available, some other means has to be found, more often than not in the form of gas cylinders. It is possible to culture cells without the use of an enhanced carbon dioxide atmosphere by the use of a zwitterionic buffer such as HEPES in a medium with a reduced sodium bicarbonate concentration.

Suppliers of industrial gases

Air Products Inc., 7201 Hamilton Boulevard, Allentown, Pennsylvania 18195-1501, USA   Tel: (610)-481-4911   Fax: (610)-481-3765

Air Products plc, Hersham Place, Molesey Road, Walton-on-Thames, Surrey KT12
    4RZ, UK   Tel: 01932 249200   Fax: 01932 249565
B.O.C. Special Gases, 24 Deer Park Road, London SW19 3UF, UK   Tel: 0181-542
    6677   Fax: 0181-543 7268
B.O.C. Group Inc., 575 Mountain Avenue, Murray Hill, New Jersey 07974-2082
    USA   Tel: (908)-665-2400   Fax: (908)-464-9015

Cryopreservation equipment

The techniques employed in cryopreservation are detailed in Chapter 7. The
equipment required to freeze cells in a viable state for storage in liquid nitrogen all employ the same principle of controlled rate cooling and freezing of
the cell suspension to minimize cell damage. Depending on the degree of
budgetary constraint, there are three types of equipment of varying
sophistication available:

The 'In-house' method

This is the original and most basic passive cell-freezing method used by many
cell culture laboratories before the development of commercial equipment.
This involves the use of a vessel (often a polystyrene box or drinking cup)
lined with several layers of laboratory tissues into which cryotubes are placed
and covered before overnight storage in a -70 °/-80 °C freezer. Tubes are
transferred the next day to liquid nitrogen. Once the correct conditions have
been established, this method can prove to be acceptably reproducible and
reliable. Relatively inexpensive commercial alternatives based on this
method are now available and have a very similar performance.

Commercial passive cell freezers

'Bicell' (Nihon Freezer Co. Ltd)   This is an insulated somewhat cylindrical
plastic vessel with a screw cap, which holds a variety of tubes or ampoules,
e.g. up to 8 × 1.8 ml cryotubes or 10 × 1.0 ml ampoules. When placed at
-80 °C, the wall of the vessel is of the correct thickness to provide the
required cooling rate.

'Mr Frosty' Cryo 1 °C freezing container (Nalgene Cat. No. 5100–0001,0002)
This is a transparent circular screw-capped plastic container which will hold
a variety of tubes or ampoules, e.g. 18 × 1.8 ml (-0001) or 12 × 5ml (-0002)
cryotubes. It employs a reservoir of isopropanol as the 'insulation' to achieve
cooling at the required rate when placed at -80 °C.

Electronic controlled-rate cell freezers.

The most commonly used models of this type of equipment are those manufactured by Planer Biomed. Their range includes models of varying degrees of sophistication and cost.

Suppliers of cryogenic equipment

Electronic controlled-rate freezers

Planer Biomed, Windmill Road, Sunbury-on-Thames, Middlesex TW16 7HD, UK
Tel: 01932 779997   Fax: 01932 781151   'Kryo' series

Passive cell freezers

Nalge Company, PO Box 20365, Rochester, New York 14602-0365, USA   Tel: (716)-586-8800   Fax: (716)-586-3294 'Nalgene' range
Nihon Freezer Company Ltd, 19-4 Yushima 3, Bunkyo-Ku, Tokyo 113, Japan   Tel: 0081-33-831-7643   Fax: 0081-33-832-0633

Ultra-low mechanical freezers ($-70°$ C or below)

Camlab Ltd, Nuffield Road, Cambridge CB4 1TH, UK   Tel: 01223 424222   Fax: 01223 420856
Denley Instruments, Life Sciences International (UK) Ltd, Unit 5, The Ringway Centre, Edison Road, Basingstoke, Hampshire RG21 2YH, UK   Tel: 01256 817282   Fax: 01256 817292
Forma Scientific Inc., Millcreek Road, PO Box 649, Marietta, Ohio 45750, USA   Tel: (614)-373-4763   Fax: (614)-373-6770
Harris Manufacturing, Puffer Hubbard Division, 275 Aiken Road, Asheville, North Carolina 28804, USA   Tel: (704)-658-2711   Fax: (704)-658-0363
Heraeus Instruments GmbH, Laboratory Division, PO Box 15 63, 63405 Hanau, Germany   Tel: +49 6181 35300   Fax: +49 6181 355973
Heto Equipment (UK) Ltd, PO Box 31, Camberley, Surrey GU15 1TN, UK
Heto-Holten A/S, Gydevang 17-19, DK 3450, Allerod, Denmark   Tel: +45 48 14 27 77   Fax: +45 48 17 46.55
Jouan Inc., 110B. Industrial Drive, Winchester, Virginia 22602, USA   Tel: (540)-869-8623   Fax: (540)-869-8626
Jouan Ltd, Merlin Way, Quarry Hill Road, Ilkeston, Derbyshire DE7 4RA, UK   Tel: 0115 944 7989   Fax: 0115 944 7080   E-mail: jouan.co.uk
Jouan SA, CP.3203, 44805 St.Herblain Cedex, France   Tel: 40 16 80.00   Fax: 40 94 70 16
Lab-Impex Research Ltd, Unit 5, Kingsway Business Park, Oldfield Road,

Hampton, Middlesex TW12 2HD, UK   Tel: 0181-296 1122   Fax: 0181-296 1123

Lab-Mart (Cambridge) Ltd, 1 Pembroke Avenue, Waterbeach, Cambridgeshire CB5 9QR, UK   Tel: 01223 861665   Fax: 01223 861990

New Brunswick Scientific Inc., 44 Talmadge Road, Edison, New Jersey 08818-4005, USA   Tel: (908)-287-1200   Fax: (908)-287-4222

New Brunswick Scientific (UK) Ltd, Edison House, 163 Dixons Hill Road, North Mymms, Hatfield, Hertfordshire AL9 7JE, UK   Tel: 01707 275733/275707   Fax: 01707 267859

Nu-Aire Inc., 2100 Fernbrook Lane, Plymouth, Minnesota 55447, USA   Tel: (612)-553-1270   Fax: (612)-553-0459

Queue Systems, 275 Aiken Road, Asheville, North Carolina 28804, USA   Tel: (704)-658-2711   Fax: (704)-658-0363

Revco Scientific, 275 Aiken Road, Asheville, North Carolina 28804, USA   Tel: (704)-658-2711   Fax: (704)-645-3368

Revco Scientific International, PO Box 321, 8600–AH Sneek (FRL), The Netherlands   Tel: 31 51557 5105   Fax: 31 51557 4659

Sanyo Electric Company Ltd, Sanyo Electric Trading Co. Ltd, 5-15 Hiyoshi-cho 2-chome, Moriguchi City, Osaka 570, Japan   Tel: (06)-992-1521   Fax: (06)-992-2874

Sanyo Gallenkamp plc, Park House, Meridian East, Meridian Business Park, Leicester LE3 2UZ, UK   Tel: 0116 2630530   Fax: 0116 2630353

Sanyo Scientific, 900 North Arlington Heights Road, Suite 310, Itasca, Illinois 07848, USA   Tel: (708)-875-3532   Fax: (708)-775-0044

So-Low Environmental Equipment Co. Inc., 10310 Spartan Drive, Cincinnati, Ohio 45215, USA   Tel: (513)-772-9410   Fax: (513)-772-0570

Liquid nitrogen

Liquid nitrogen suppliers

Air Products Inc., 7201 Hamilton Boulevard, Allentown, Pennsylvania 18195-1501, USA   Tel: (610)-481-4911   Fax: (610)-481-3765

Air Products plc., Hersham Place, Molesey Road, Walton-on-Thames, Surrey KT12 4RZ, UK   Tel: 01932 249200   Fax: 01932 249565

BOC Cryoplants Ltd, Angel Road, London, N18 3BW, UK   Tel: 0181-803 1300   Fax: 0181-884 1389

BOC Group Inc., 575 Mountain Avenue, Murray Hill, New Jersey 07974-2082, USA   Tel: (908)-665-2400   Fax: (908)-464-9015

Cryoservice Ltd, Blackpole Trading Estate, Blackpole Road, Worcester WR3 8SG, UK   Tel: 01905 754500   Fax: 01905 754060

Liquid nitrogen refrigerator suppliers

Barnstead-Thermolyne Corporation, 2555 Kerper Boulevard, Dubuque, Iowa 52001, USA   Tel: (319)-556-2241   Fax: (319)-556-0695

International Cryogenics Inc., 4040 Championship Drive, Indianapolis, Indiana 46268, USA   Tel: (317)-297-4777   Fax: (317)-297-7988

L'Air Liquide, Division Materiel Cryogenique, Parc Gustave-Eiffel, 8 rue Gutenberg, Bussy-Saint-Georges, 77607 Marne-La-Valle, Cedex 3, France Tel; 64 76 15 00   Fax: 64 76 16 99

MVE Cryogenics, 8011 34th. Avenue South, Suite 100, Bloomington, Minnesota 55425-1636, USA   Tel: (612)-853-9666   Fax: (612)-853-9661

Statebourne Cryogenics Ltd, 18 Parsons Road, Washington, Tyne and Wear NE37 1EZ, UK   Tel: 0191-416 4104   Fax: 0191-415 0369

Taylor-Wharton Cryogenics, 4075 Hamilton Boulevard, Theodore, Alabama 36590–0568, USA   Tel: (334)-443-2209   Fax: (334)-443-8680

Freezer storage equipment

Nalge (Europe) Ltd, Foxwood Court, Rotherwas, Hereford HR2 6JQ, UK   Tel: 01432 263933   Fax: 01432 351923

Nalge Nunc International, 75 Panorama Creek Drive, PO Box 20365, Rochester, New York, 14602-0365, USA   Tel: (716)-264-3898   Fax:(716)-264-3706 'Nalgene' Ware

Roller apparatus

Where larger amounts of cells are required, these can be grown in roller bottles. Machines for rotating roller vessels, suitable for culturing either adherent or suspension cells, are available usually in modular form. These have a base deck containing the drive motor mechanism and places for two vessels per deck (for use in an incubator) or five places (for use in a warm room) to which additional decks can be added. The drive is usually variable speed.

Manufacturers of roller equipment for cell culture bottles

Lab-Line Instruments Inc., 15th and Bloomingdale Avenues, Melrose Park, Illinois 60160, USA   Tel: (708)-450–2600   Fax: (708)-450–0943   'Cel-Gro'

Matrix Technologies Corporation, 44 Stedman Street, Lowell, Massachusetts 01851-2734, USA   Tel: (508)-454-5690   Fax: (508)-458-9174   Tecnomara 'Cel-Roll'

Wheaton Ltd, Division of Wheaton Industries, 1301 North Tenth Street, Millville, New Jersey 08332, USA   Tel: (609)-825-1100   Fax: (609)-825-1368

Spinner vessels

These can be used to grow suspension cells, or adherent cells using micro-carriers. Microcarriers are small porous or solid beads that essentially increase the effective surface area available for adherence of monolayer cultures (see Chapter 9). Obtainable in various sizes up to 36 litres capacity with variable numbers of access ports and a range of paddle designs. Used in conjunction with a spinner base (magnetic stirrer) or an overhead drive depending on the manufacturer and the model.

Manufacturers of spinner apparatus

Bellco Glass Inc., 340 Erudo Road, Vineland, New Jersey 08360, USA   Tel: (609)-691-1075   Fax: (609)-691-3247

Camlab Ltd, Nuffield Road, Cambridge CB4 1TH, UK   Tel: 01223 424222   Fax: 01223 420856

Cellon Sarl, 204 route d'Arion, L-8010 Straspen, Luxembourg   Tel: +352 312 313   Fax: +352 311 052

H & P Labortechnik GmbH, Bruckmannring 28, D-85764 Oberschelsshheim, Munich, Germany   Tel: +49 89 315 8220

Integra Biosciences AG, PO Box 74, CH-8304, Wallisellen, Switzerland   Tel: 830 2277   Fax: 830 7852

Integra Biosciences Inc., 10–11 Roessler Road, Woburn, Massachusetts 01801, USA   Tel: (617)-937-8300   Fax: (617)-937-8301

Integra Biosciences Ltd, The Annexe, New Barnes Mill, Cottonmill Lane, St Albans, Hertfordshire AL1 2HB, UK   Tel: 01727 848825   Fax: 01727 811769 'Cellspin"

Lab-Line Instruments Inc., 15th and Bloomingdale Avenues, Melrose Park, Illinois 60160–1491, USA   Tel: (708)-450–2600   Fax: (708)-450–0943

Techne (Cambridge) Ltd, Duxford, Cambridge CB2 4PZ, UK   Tel: 01223 832401   Fax: 01223 836838

Techne Inc., 743 Alexander Road, Princeton, New Jersey 08540–6328, USA   Tel: (609)-452-9275   Fax: (609)-987-8177

Wheaton Ltd, Division of Wheaton Industries, 1301 North Tenth Street, Milville, New Jersey 08332, USA   Tel: (609)-825-1100   Fax: (609)-825-1368

Media handling equipment

Water purification equipment

Many laboratories use distilled or deionized water for a variety of purposes including the preparation of cell culture media and reagents. The minimum quality needed for successful cell culture is water with a resistance of at least 1 megohm (=1.0 uS/cm conductivity) or better. This is equivalent to double distilled water classified as Grade 1 by the College of American Pathologists (CAP). Some laboratories still use stills which can be useful where only small volumes of water are required, although with current energy costs this method is rather expensive. Where stills are used, they need to be made of glass to minimize trace ion contamination and need cleaning out on a regular basis. Most laboratories now use deionizers and these vary considerably in capacity. Modular cartridge systems are available where larger volumes of water are required. These comprise a prefilter, organic scavenger, deionizer, depyrogenator and a final filter to 0.22 μM. Meters for testing the conductivity and/or resistivity of water are widely available.

Manufacturers of water purification and testing equipment

Water conductivity meters

ATI Russell, ATI Unicam, York Street, Cambridge CB1 2PX, UK   Tel: 01223 358866   Fax: 01223 31276

Hanna Instruments Inc., 584 Park East Drive, Woonsocket, Rhode Island 02895, USA   Tel: (401)-765-7500   Fax: (401)-766-3087

Hanna Instruments Ltd, Eden Way, Pages Industrial Park, Leighton Buzzard, Bedfordshire LU7 8TZ, UK   Tel: 01525 850855   Fax: 01525 853668

Jenway Ltd, Gransmere Green, Felsted, Dunmow, Essex CM6 3LB, UK   Tel: 01371 820122   Fax: 01371 821083

Knick Elektronische Messegeraete, Beuckestrasse 22, 1000 Berlin 37, Germany   Tel: 0049.80010   Fax: 0049.8001635

Radiometer Analytical SA, 72 rue d'Alsace, F-69627, Villeurbanne Cedex, Lyons, France   Tel: 78.03.38.38   Fax: 78.68.88.12

Radiometer Ltd, Manor Court, Manor Royal, Crawley, West Sussex RH10 2BR, UK   Tel: 01293 517599   Fax: 01293 531597

WPA, The Old Station, Linton, Cambridge CB1 6NW, UK   Tel: 01223 892668   Fax: 01223 894118

WTW GmbH, Dr Karl-Slevogt-Strasse 1, D-82362 Weilheim, Germany   Tel: 0881.1830   Fax: 0881.69529

YSI Inc., 1725 Brannum Lane, PO Box 279, Yellow Springs, Ohio 45387, USA   Tel: (513)-767-7241   Fax: (513)-767-9353

Water distillation equipment

Bibby Sterilin Ltd, Tilling Drive, Stone, Staffordshire ST15 0SA, UK   Tel: 01785 812121   Fax: 01785 813748   'Distinction' range

Sanyo Gallenkamp plc, Park House, Meridian East, Meridian Business Park, Leicester LE3 2UZ, UK   Tel: 0116 2630530   Fax: 0116 2630353   'Fi-Streem' range

Vaponics Ltd, 20 Park Street, Princes Risborough, Buckinghamshire HP27 9AH, UK   Tel: 01844 346811   Fax: 01844 274216

Water purification equipment (deionizers)

J.Bibby Science Products Ltd, Tilling Drive, Stone, Staffordshire ST15 0SA, UK   Tel: 01785 812121   Fax: 01785 813748   'Aquatron' range

Elga Ltd, High Street, Lane End, High Wycombe, Buckinghamshire HP14 3JH, UK   Tel: 01494 881393   Fax: 01494 881007

Elga Inc., 430 Old Boston Road, Topsfield, Massachusetts 01983, USA   Tel: (508)-887-6300   Fax: (508)-887-6266

Millipore Corporation Inc., 80 Ashby Road, Bedford, Massachusetts 01730, USA   Tel: (617)-275-9200   Fax: (617)-275-5550

Millipore Europe, B.P.307, Saint-Quentin-en-Yvelines, F-78054 Cedex, France   Tel: 33(1)30 12 7000   Fax: 33(1)30 12 7182

Millipore (UK) Ltd, The Boulevard, Blackmoor Lane, Watford, Hertfordshire WD1 8YW, UK   Tel: 01923 816375   Fax: 01923 818295

Purite Ltd, Bandet Way, Thame, Oxfordshire OX9 3SJ, UK   Tel: 01844 217141   Fax: 01844 218098

Sation, Luchana, 77-08005 Barcelona, Spain   Tel: 00 34 93 300 7513   Fax: 00 34 93 309 3364

US Filter/Ionpure, 10 Technology Drive, Lowell, Massachusetts 01851, USA   Tel: (508)-934-9349   Toll Free 800–875 PURE   Fax: (508)-453-5821/(508)-441-6025

USF Ltd, Harforde Court, Foxholes Business Park, John Tate Road, Hertford SG13 7NW, UK   Tel: 01992 823300   Fax: 01992 501528

USF Liquipure, Industriegebiet Struth, Eulerstrasse, 56235 Rambach-Baumbach, Germany   Tel: +49 26 23 8910   Fax: +49 26 23 891220

Filters

Where reagents and media are prepared in-house, they will usually be sterilized by filtration. A wide range of filtration equipment is available, and the type and size used will depend on the volumes to be filtered. These range from small housings that fit on the end of a syringe for filtering just a few millilitres to large, flat membrane or cartridge assemblies for larger volumes. Both reusable and disposable assemblies are available. In every case, the requirement is to remove microbiological contaminants for which the maximum pore size of the filter should be 0.22 µM. Other filters of a smaller pore size are also available to allow the removal of mycoplasma, e.g. Pall N.66 Posidyne or Gelman Supor DCF and viruses, e.g. Pall Ultipor VF Grade DV50 or Gelman Hi-Flo PN SCF92 HP61. It is recommended that, when filtering cell culture medium, the final filtration should be through a 0.10 μM membrane to ensure the removal of both bacteria and mycoplasma. For volumes up to 1 litre, a 47 mm membrane is ideal. For more viscous solutions, e.g. serum, a series of coarser clarifying filters will need to be used before the final 0.10 μM filtration.

Manufacturers of filtration equipment

Becton Dickinson UK Ltd, Between Towns Road, Cowley, Oxford OX4 3LY, UK
 Tel: 01865 777722   Fax: 01865 717313   Vacu-Cap

Becton Dickinson Europe, 5 Chemin des Sources, 38241 Meylan Cedex, France
 Tel: +33 76 41 64 64   Fax: +33 76 90 19 65

Becton Dickinson Canada Inc., 2464 South Sheridan Way, Mississauga, Ontario L5J
 2M8, Canada   Tel: (905)-822-4820   Fax: (905)-822-2644

Gelman Sciences, 600 South Wagner Road, Ann Arbor, Michigan 48103-9019,
 USA   Tel: (313)-665-0651   Fax: (313)-913-6114/6197

Gelman Sciences Ltd, Brackmills Business Park, Caswell Road, Northampton
 NN14 7EZ, UK   Tel: 01604 765141   Fax: 01604 761383

Millipore Corporation Inc., 80 Ashby Road, Bedford, Massachusetts 01730, USA
 Tel: (617)-275-9200   Fax: (617)-275-5550

Millipore (UK) Ltd, The Boulevard, Blackmoor Lane, Watford, Hertfordshire WD1
 8YW, UK   Tel: 01923 816375   Fax: 01923 818295

Nippon Becton Dickinson Company Ltd, Shimato Building, 5-34, Akasaka-8-
 chome, Minato-ku, Tokyo 107, Japan   Tel: (81)-3.3403-9991   Fax: (81)-
 3.3403-5321

Pall Filtron Corporation, 50 Bearfoot Road, Northborough, Massachusetts 01532,
 USA   Tel: Toll Free 800–FILTRON   Fax: (508)-393-1874

Pall Ultrafine Group, Europa House, Havant Street, Portsmouth PO1 3PD, UK
 Tel: 01705 303303   Fax: 01705 302506

Sartorius Corporation, 131 Heartland Boulevard, Edgewood, New York 95054,
 USA   Tel: (516)-254-4249   Fax: (516)-254-4253

Sartorius GmbH, Postfach 3243, Weender Landstrasse 94-108, 3400 Gottingen,
 Germany   Tel: 0551.3080   Fax: 0551.380289

Sartorius Ltd, Longmead Business Centre, Blenheim Road, Epsom, Surrey KT19
 9QN, UK   Tel: 01372 745811   Fax: 01372 720799

Schleicher and Schuell GmbH, PO Box 4, D-3354 Dassel, Germany   Tel:
 05561.7910   Fax: 05564.2309

Schleicher and Schuell Inc., 10 Optical Avenue, PO Box 2012, Keene, New
 Hampshire 03431, USA   Tel: (603)-352-3810   Fax: (603)-357-3627

Whatman International Ltd, Whatman House, St.Leonards Road, 20/20 Maidstone, Kent ME16 0LJ, UK   Tel: 01622 676670   Fax: 01622 677011

Whatman Inc., 260 Neck Road, Haverhill, Massachusetts 01835-0723, USA   Tel: (508)-374-7400   Fax: (508)-374-7070

Osmometer

It is important that all cell culture medium produced in-house is tested for osmolarity before use to check that each batch comes within the accepted limits for that particular type of medium (see Chapter 3, Table 6). A simple way of doing this is with an osmometer. These are available from many laboratory suppliers and are relatively inexpensive.

Manufacturers of osmometers

Advanced Instruments Inc., Two Technology Way, Norwood, Massachusetts 02062, USA   Tel: (617)-320–9000   Fax: (617)-320–8181 Model 65-31

Firma Hermann Roebling, Kattewag 32, D-14129 Berlin 38, Germany   Tel: +49 3080 35671   Fax: +49 3080 36252

Fiske Associates Inc., Quaker Highway, Uxbridge, Massachusetts 01569, USA   Tel: (617)-278-2482

Miscellaneous equipment

Cell scrapers

Available from some of the manufacturers of disposable plasticware, e.g. Falcon. Useful for the removal of adherent cells from the surface of flat culture vessels without the use of proteolytic enzymes such as trypsin. These can also be readily made using silicone rubber bungs that have been cut in half and attached to the end of a 0.2 ml plastic disposable pipette.

Pipettes

Pipettes are an essential part of cell culture. They are available either as disposable, manufactured in glass or plastic, or re-usable and manufactured in glass. Most laboratories carry a range of graduated pipettes of various volumes from 0.2 ml through 1 ml, 2 ml, 5 ml, 10 ml to 25 ml as well as Pasteur pipettes. Disposable pipettes come pre-packed and sterile ready-for-use: re-usable pipettes require disinfecting after use, washing, replugging with non-absorbent cotton wool, repacking (usually in aluminium cans) and dry heat sterilizing, e.g. in a dry oven at 180 °C. *Never* use an unplugged pipette for cell culture work (except with a media pump), even if the pipette aid has a filter on it. Pipettes are obtainable from any good laboratory supplier.

Pipette aids

Because mouth pipetting is strictly forbidden, some form of pipette aid is essential for filling and emptying pipettes. Many different types and makes are available from most good laboratory suppliers. These range from the simple plastic/rubber bulb to the motorized pump type with various hand-operated models in-between.

Gas burners

The use of gas burners in laminar flow hoods is not recommended because gas burners cause serious disruption to the airflow within the cabinet, reducing the effective protection of the laminar flow air curtain. They can also cause damage to the structure of the cabinet and cannot be used in conjunction with plasticware. Where the use of a gas burner is considered essential, or when using glass bottles on the open bench, a gas burner of the Fireboy type (Integra Biosciences) is recommended. This is operated either by a hand sensor, foot pedal or timer and will only remain alight for as long as required.

Supplier of Fireboy gas burner

Integra Biosciences AG, PO Box 74, CH-8304, Wallisellen, Switzerland.   Tel: 830 2277   Fax: 830 7852

Integra Biosciences Inc., 10–11 Roessler Road, Woburn, Massachusetts 01801, USA Tel: (617)-937-8300   Fax: (617)-937-8301

Integra Biosciences Ltd, The Annexe, New Barnes Mill, Cottonmill Lane, St Albans, Hertfordshire AL1 2HB, UK   Tel: 01727 848825   Fax: 01727 811769

Plastic tubes and bottles

These are obtainable in several different designs and volumes. Some can be sterilized by autoclaving and some are disposable. They are used most frequently to sediment (pellet) adherent cells during the process of subculturing (passaging) or during the preparation of both adherent and suspension cells for storage in liquid nitrogen. The most commonly used are the Universal bottle with a working volume of approximately 22 ml and the 50 ml centrifuge tube. Both these have conical bottoms: the Universal has a skirt at the bottom to allow it to be free-standing; the 50 ml tube is available with or without a bottom skirt. These are all disposable. For larger volumes there is the 175 ml conical centrifuge tube and centrifuge bottles of varying sizes from 250 ml to 1 litre all of which are re-usable and steam sterilizable.

Manufacturers of plasticware suitable for cell culture

Becton Dickinson Europe, Headquarters BP No.37-38241, Meylan Cedex, France
Tel: 33 76 90 80 35   Fax: +33 76 90 19 65   'Falcon' brand

Becton Dickinson Labware, Two Oak Park, Bedford, Massachusetts 01730, USA
Tel: (617)-275-0004   Fax: (617)-275-0043

Becton Dickinson (UK) Ltd, Between Towns Road, Cowley, Oxford OX4 3LY, UK
Tel: 01865 777722   Fax: 01865 717313

Bibby Sterilin Ltd, Tilling Drive, Stone, Staffordshire ST15 0SA, UK   Tel: 01785 812121   Fax: 01785 813748

Corning brand range: obtainable from Bibby Sterilin Ltd in UK

Corning Costar Corporation, One Alewife Center, Cambridge, Massachusetts 01240, USA   Tel: (617)-868-6200   Fax: (617)-868-2076

Costar Europe Ltd, PO Box 94, 1170 AB Badhoevedorp, The Netherlands   Tel: 020 659 6051   Fax: 020 659 7673

Costar UK Ltd, 10 The Valley Centre, Gordon Road, High Wycombe, Buckinghamshire HP13 6EQ, UK   Tel: 01494 471207   Fax: 01494 464891

Greiner GmbH, Maybachstrasse 2, D-72636 Frickenhausen, Germany   Tel: 07 022 50 10   Fax: 07 022 50 15 14

Greiner Labortechnik Ltd, Station Road, Cam, Dursley, Gloucester GL11 5NS, UK
Tel: 01453 548833   Fax: 01453 544070

Nalge Nunc International, PO Box 20365, 75 Panorama Creek Drive, Rochester, New York 14602, USA   Tel: (716)-586-8800   Fax: (716)-586-8431

Nunc A/S, Kamstrupvej 90, PO Box 280, DK-4000 Roskilde, Denmark   Tel: (+45) 46 35 90 65   Fax: (+45) 46 35 01 05   E-mail: infociety@nunc.dk

Nunc A/S, Life Technologies Ltd, 3 Fountain Drive, Inchinnan Business Park, Paisley PA4 9RF, Scotland, UK   Tel: 0141-814 6100   Fax: 0141-887 1167

Nunc Inc., 2000 North Aurora Road, Naperville, Illinois 60563-1796, USA   Tel: (708)-983-5700   Fax: (708)-416-2556

Sterilin brand range, see Bibby Sterilin Ltd

Media pumps

In an attempt to reduce the risk of chance contamination of their cell cultures, some workers prefer not to pour medium into, and out of, flasks or

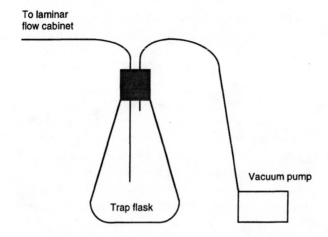

Fig. 2.2. Media pump set-up. See text for description of operation.

use a pipette for decanting medium from Petri dishes but elect instead to use a media pump to remove the medium. These are usually constructed in-house (see Fig. 2.2) and consist of a vacuum source (either a vane or water pump) connected to a 2- or 5-litre flask which acts as a trap. The flask has a rubber bung bored to take two lengths of 7 mm ID glass or rigid plastic tube, one of which is sufficiently long to reach half way down the flask and the second shorter length to reach just below the bung. Both pieces of glass/plastic tube protrude approximately 2 cm above the top of the bung. The pump is connected to the shorter piece of glass/plastic tube by a length of thick-walled silicone rubber tubing containing an airline filter. Both the pump and the trap are external to the laminar flow cabinet. To the longer piece of glass tube is connected a length of 6 mm ID silicone rubber tubing which must be long enough to stretch to well within the laminar flow cabinet. For each cell type or individual operation a sterile Pasteur pipette is attached to the end of the tubing inside the cabinet after switching on the vacuum pump. The trap flask can contain a small amount of suitable disinfectant but must be disconnected and washed thoroughly at the end of each day it has been in use. Some laboratories use surgical suction pumps bought commercially for this purpose.

Suppliers of surgical suction pumps

Aerosol Medical, Wyncolls Road, Colchester, Essex CO4 4HT, UK   Tel: 01206 842244   Fax: 01206 845849

Rocialle Medical Ltd, 19 Clifton Road, Cambridge CB1 4WY, UK   Tel: 01223
    242425   Fax: 01223 215346

Water baths

Widely available from laboratory suppliers, it is important to note that water
baths can play a significant role in success or failure in cell culture and too
often are subject to varying degrees of neglect. It is essential that any water
bath used for thawing reagents or warming media is kept clean because even
apparently clear water can harbour a significant amount of microbiological
contaminants. Various approaches have been adopted to overcome this
problem: the best way is for the bath to be emptied and cleaned at the end
of each week and refilled on a Monday morning with fresh deionized or distilled water. Some laboratories add non-corrosive disinfectants or quaternary
ammonium compounds such as cetyltrimethyl ammonium bromide
('Cetrimide') to the bath to prevent the build-up of micro-organisms. It is
preferable to use a bath without a lid to reduce the amount of condensation
around the bottle neck particularly. Polypropylene insulating spheres can be
purchased that float on the surface of the bath and reduce the level of water
evaporation. Always dry the outside of every bottle thoroughly immediately
after removing it from the water bath. A satisfactory alternative to the water
bath is the heater block or hot plate which, although slower, is less likely to
be a source of microbiological contamination. These are obtainable from
most laboratory suppliers.

3

Cell culture media

Choice of a suitable culture medium, appropriate for the type of cell it is intended to cultivate, is very important. For established cell lines, the medium used will have been pre-determined and it is then usually only a question of continuing with the same medium, serum and culture conditions. Over the history of cell culture, a large number of different types of medium and balanced salt solutions has been formulated and published but even today the vast majority of cells are cultivated in one of seven or eight different media. The formulations for six of the more popular ones are given in Table 3.1 and discussed in this chapter.

Many of the other formulations were developed specifically for the cultivation of cells with more specialized nutritional requirements, and most commercial suppliers offer a selection of both types of medium. For primary cell cultures, the type of medium used is the choice of the researcher. More often than not, an established cell line can be adapted to grow in an alternative medium should the need arise but, when doing so, cultures should always be run in parallel in order to monitor the comparison until adaption has been assessed to have fully occurred.

Sources

How the researcher obtains their medium will usually depend on the type of organization in which they carry out their research. Centralized in-house production or a Stores stock system will usually mean ready access to medium but, in all other cases, the medium will either have to be purchased directly from a commercial supplier or made in the laboratory. The amounts purchased or made will depend on the number of cells required and the frequency. Commercial media can usually be purchased either as a 1 × liquid, a 10 × concentrated liquid or as powder. It is also possible to purchase some

Table 3.1. *Formulations of the most commonly used types of cell culture media*

Component	BME[1] (mg/l)	MEM[2] (mg/l)	DMEM[3] (mg/l)	RPMI[4] (mg/l)	McCoys 5A[5] (mg/l)	Hams F.12[6] (mg/l)
Inorganic salts						
$CaCl_2$ anhydrous	200.00	200.00	200.00	–	100.00	33.20
$Ca(NO_3)_2 \cdot 4H_2O$	–	–	–	100.00	–	–
KCl	400.00	400.00	400.00	400.00	400.00	223.60
$Fe(NO_3)_3 \cdot 9H_2O$	–	–	0.10	–	–	–
$FeSO_4 \cdot 7H_2O$	–	–	–	–	–	0.834
$MgCl_2$ anhydrous	–	–	–	–	–	57.22
$MgSO_4 \cdot 7H_2O$	200.00	200.00	200.00	100.00	200.00	–
$NaCl$	6800.00	6800.00	6400.00	6000.00	6460.00	7599.00
$NaHCO_3$	2200.00	2200.00	3700.00	2000.00	2200.00	1176.00
Na_2HPO_4 anhydrous	–	–	–	800.00	–	141.98
$NaH_2PO_4 \cdot H_2O$	140.00	140.00	125.00	–	580.00	–
$ZnSO_4 \cdot 7H_2O$	–	–	–	–	–	0.863
Other components						
D-Glucose	1000.00	1000.00	4500.00	2000.00	3000.00	1801.600
Phenol Red	10.00	10.00	15.90	5.30	11.00	12.0
Sodium pyruvate	–	–	110.00	–	–	110.10
Glutathione (reduced)	–	–	–	1.00	0.50	–
Hypoxanthine	–	–	–	–	–	4.08
Linoleic acid	–	–	–	–	–	0.084
DL-6,8 Thioctic acid	–	–	–	–	–	0.20
Putrescine 2HCl	–	–	–	–	–	0.1611
Thymidine	–	–	–	–	–	0.7266
HEPES(d)	(5958.00)	(5958.00)	(5958.00)	(5958.00)	(5960.00)	–

Table 3.1 (*cont.*)

Amino acids

L-alanine	–	–	–	–	13.36	8.91
L-arginine HCl	21.06	126.40	84.00	200.00[a]	42.14	210.70
L-asparagine	–	–	–	56.82	45.03	13.00
L-aspartic acid	–	–	–	20.00	19.97	13.30
L-cysteine	12.01	24.02	–	–	24.24	36.00[b]
L-cystine	–	–	48.00	50.00	–	–
L-glutamic acid	–	–	–	20.00	22.07	14.71
L-glutamine	292.30	292.30	584.00	300.00	219.15	146.20
Glycine	–	–	30.00	10.00	7.51	7.51
L-histidine	8.00	41.90[c]	42.00[c]	15.00[a]	20.96[c]	20.96[c]
L-hydroxyproline	–	–	–	20.00	19.67	–
L-isoleucine	26.23	52.50	104.80	50.00	39.36	3.94
L-leucine	26.23	52.50	104.80	50.00	39.36	13.12
L-lysine HCl	36.53	73.06	146.20	40.00	36.54	36.54
L-methionine	7.46	14.90	30.00	15.00	14.92	4.48
L-phenylalanine	16.51	33.02	66.00	15.00	16.52	4.96
L-proline	–	–	–	20.00	17.27	34.53
L-serine	–	–	42.00	30.00	26.28	10.51
L-threonine	23.82	47.64	95.20	20.00	17.87	11.91
L-tryptophan	4.08	10.20	16.00	5.00	3.06	2.04
L-tyrosine	18.11	36.22	72.00	20.00	18.12	5.44
L-valine	23.43	46.90	93.60	20.00	17.57	11.71

Vitamins

Ascorbic acid	–	–	–	–	0.56	–
D-biotin	1.00	–	–	0.20	0.20	0.0073
D-Ca pantothenate	1.00	1.00	4.00	0.25	0.20	0.48

Table 3.1 (*cont.*)

Component	BME[1] (mg/l)	MEM[2] (mg/l)	DMEM[3] (mg/l)	RPMI[4] (mg/l)	McCoys (mg/l)	5A[5] Hams F12[6] (mg/l)
Choline chloride	1.00	1.00	4.00	3.00	5.00	13.96
Folic acid	1.00	1.00	4.00	1.00	10.00	1.324
i-inositol	1.80	2.00	7.20	35.00	36.00	18.02
Nicotinamide	1.00	1.00	4.00	1.00	0.50	0.037
Nicotinic acid	–	–	–	–	0.50	–
PABA	–	–	–	1.00	1.00	–
Pyridoxal HCl	1.00	1.00	4.00	1.00	0.50	–
Pyridoxine HCl	–	–	–	–	0.50	0.062
Riboflavin	0.10	0.10	0.40	0.20	0.20	0.0376
Thiamine HCl	1.00	1.00	4.00	1.00	0.20	0.337
Vitamin B$_{12}$	–	–	–	0.005	2.00	1.357

Notes:
[a] free base. [b] HCl. [c] HCl·H$_2$O. [d] optional.

Sources:

[1] Eagle (1955, 1956a, 1956b).

[2] Eagle (1959)

[3] Dulbecco & Freeman (1959). Smith, Freeman, Vogt & Dulbecco (1960).

[4] Moore, Gerner & Franklin (1967).

[5] Neuman and McCoy (1958). Growth–promoting properties of pyruvate, oxalacetate and α–ketoglutarate for isolated Walker Carcinosarcoma 256 cells. McCoy, Maxwell & Kruse. (1959).

[6] Ham (1965).

types of the more common media as a powder that, after solution, can be autoclaved rather than filter sterilized. To perform successful cell culture will require a range of reagents and solutions in addition to the culture medium, all of which can be made in-house or purchased from suppliers.

Functions of the main ingredients of cell culture media

Balanced salt solution (BSS)

The BSS is designed to provide a selection of metabolically important inorganic salts in concentrations sufficient not to reach depletion during the passage period of the cell culture and balanced to provide a combined osmotic pressure similar to that experienced by the cell type *in vivo*. It also contains sodium bicarbonate which, in most cases, acts primarily as a buffer but also functions as an important metabolite. The phosphates also function as a secondary source of buffering.

Amino acids and vitamins

These solutions provide the selection of essential nutrients required by mammalian cells for growth and division.

Other ingredients

The selection of the other ingredients varies according to the type of medium which, in turn, will depend on the type of cell and the function for which they are being grown. They will always include a prime source of energy, usually in the form of glucose, and an indicator to aid visual assessment of the pH of the medium, usually Phenol Red.

Buffers

Cell metabolism produces comparatively large amounts of waste products that are usually acidic in nature. To counteract the rapid change in the pH of the medium that these by-products produce, all cell culture medium include at least one chemical compound whose primary function is to act as a buffer by reacting with, and to some extent neutralizing, the acidity. Secondary buffering is supplied by the phosphate salts in the balanced salt solution and to a lesser extent by any serum present.

The 'ideal' buffer is expected to exhibit the following properties:

1 consistently to maintain a given pH in a particular medium;
2 to show no interference with chemical or biochemical processes occurring in the medium;
3 to show no impedance of measurements or observations made on the
system.

The most widely used system for buffering cell culture media has tradition-
ally been sodium bicarbonate/carbon dioxide:

$$NaHCO_3+H_2O$$
$$\rightarrow Na^+ + HCO_{3_-} + H_2O$$
$$\rightarrow Na^+ + H_2CO_3 + OH^-$$
$$\rightarrow Na^+ + OH^- + H_2O + CO_2 \rightarrow increased\ alkalinity$$

Sodium bicarbonate serves three functions:

1 to maintain pH;
2 to maintain osmotic pressure;
3 to provide a source of energy.

There are two main disadvantages to the use of sodium bicarbonate:

1 it has a pK_a of 6.3 at 37 °C which gives suboptimal buffering in the pH
7.0–7.5 range;
2 it requires a special gas mixture of 5% or 10% CO_2 in 95% or 90% air in
which to equilibrate.

Various means have been devised to overcome the equilibration problem.
Hanks formulated his balanced salt solution to contain a lower concentration of sodium bicarbonate to eliminate the need for a special gas mixture.
Hanks' BSS is not as efficient as Earle's which still remains the more popular.
   Another alternative to bicarbonate as a buffer was Tris (Tris(hydroxy-
methyl) amino-methane). However, this buffer also suffers from several disadvantages:

1 it has a high chemical reactivity as a primary amine;
2 it often acts as an inhibitor;
3 it penetrates biological membranes;
4 it has a high coefficient of pH dependence with temperature.

The introduction by Good (1966) of a series of zwitterionic buffers covering the pH range from 6.15 (MES) to 9.55 (CHES) includes the buffer with
the acronym HEPES (see further under Buffers). This buffer has a pK_a of
7.31 at 37 °C which is optimal for cell culture media, does not cross the cell
membrane and equilibrates with air.

Antibiotics

Included routinely in most cell culture media as a means of reducing the incidence of opportunistic contamination by micro-organisms. The most frequent cause of contamination is usually a failure of aseptic technique combined with too much confidence being placed in the effectiveness of antibiotics. However, it is quite possible to culture mammalian cells for long periods of time in the absence of antibiotics particularly if using a laminar flow hood. The most widely used antibiotic supplement is a mixture containing penicillin and streptomycin. There are potential problems with using this mixture continuously with any cell line that is maintained in culture for long periods of time. Occasionally, gradual resistance may build up from organisms that have been introduced into the culture accidentally which may result in eventual obvious contamination or remain as a low level infection of partially resistant organisms. For this reason it is often a good idea periodically to try a test culture in antibiotic-free medium to check that there is no background contamination and also to culture the cells for a passage or two in medium containing a different antibiotic.

Laboratory-scale production

In-house production of cell culture medium is only economically viable either where there is no alternative or where the level of usage is sufficiently high to accrue economies of scale. Most cell culture media consist of four main groups of ingredients: a balanced salt solution; amino acids; vitamins; and other ingredients. If medium is prepared on a fairly regular basis, it is worth preparing concentrated stock solutions of the amino acids, vitamins, glutamine, Phenol Red and certain of the salts used in the BSS. The following is a protocol for the small-scale in-house production of the medium and reagents required for basic cell culture.

Equipment

The exact type and size of individual pieces of equipment used will depend on the size of the batch of medium being prepared. The following are the general types of equipment needed:

Preparation vessel in which to mix the ingredients (usually beaker, flask or tank[1])

Stirrer (magnetic for smaller batches, plastic rod or overhead for larger batches)

Source of purified water (see further under Quality Control)

Sterile storage bottles in which to dispense the completed medium
Sterile filter housing/cartridge(s) with ancillary lines and clamps
Sterile dispensing bell[2] (for larger batches).
Retort stands to hold dispensing bell and filter cartridges
Laminar flow cabinet (horizontal or vertical flow)
pH meter and calibration standards
1.0 M NaOH and 1.0 M HCl for pH adjustment
Supply of carbon dioxide gas for pH adjustment.
Vacuum line or pump (water or mechanical) (for smaller batches)
Pump, e.g. peristaltic or positive pressure (for larger batches)
FMM and SBC sterility test media (see under quality control)
Water bath to thaw stock aliquots
−20 °C freezer to store stock aliquots
Hydrophobic airline filter.

Notes

[1] Stainless steel tanks of varying sizes are available from Millipore and Sartorius. Plastic tanks are available from Nalgene.

[2] Reusable glass bells are available from various suppliers (see Fig. 3.1). Choose one that has a tap to allow control of the liquid flow. Much more preferable is the use of plastic disposable bells incorporating a filter assembly in the same housing, e.g Millipore Millipak series, Sartorius Sartolab P20, Gelman SpiralCap or Becton Dickinson VacuCap. This eliminates the need for a separate filter flask, membranes and housing but must still have a tap fitted into the delivery line .

Assembling the equipment for sterilization

For smaller batches of medium (up to 5 litres)

Use a 47 mm filter holder and an Erlenmeyer side-arm filtering flask. Attach the base of the filter holder to the flask by the use of a silicone (or other non-toxic rubber) bung. Place a membrane of the correct porosity on to the base. Check that the membrane is undamaged. Always handle the membrane very carefully with a pair of forceps specifically designed for picking them up. Ensure the membrane is placed centrally on to the base and not to one side. Attach the funnel (reservoir) carefully on to the base, making sure the membrane is not dislodged, and clamp down. Fit the arm with a length of silicone rubber tubing containing or ending in a 0.2 μm hydrophobic airline filter and an adjustable tubing clamp. Wrap the entire top of the flask includ-

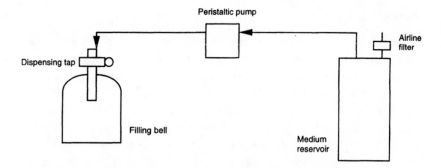

Fig. 3.1. Schematic of a recyclable glass filling-bell apparatus. The bell and reservoir are connected by a length of silicone rubber tubing suitable for use with a peristaltic pump. The bell is totally enclosed in wrapping for sterilization. See text for full explanation. A filling bell with an integral filter is sometimes referred to as a filter bell by some manufacturers.

ing the side arm tube in aluminium foil, or waterproof brown paper ('Kraft' paper), to below the arm and tape round with autoclave tape. Sterilize by autoclaving at 20 lb for 20 minutes.

For larger batches of medium

Use a mixing tank and filter bell. Assemble the filter cartridges or large filter housing (293 mm) as in the makers instructions, and attach silicone rubber tubing to the inlet of sufficient length to reach the mixing tank. If using a peristaltic pump between the tank and the filter housing ensure the tubing is suitable for this purpose. To the outlet, attach another length of tubing of sufficient length to reach the dispensing bell in its retort stand (if using a reusable glass dispensing bell attach this to the end and include a dispensing tap if necessary). Close both ends of the system with covers of aluminium foil or Kraft paper and tape round with autoclave tape. Sterilize by autoclaving at 20 lb for 20 minutes. With cartridges in pairs, an integrity test can be performed after sterilization but before use.

Preparation of the ingredients

All the ingredients listed below can be obtained from one or more of the chemical companies listed at the end of the chapter.

Amino acids, 100 ×

Although some of the amino acids are only poorly soluble in water, most of the amino acids are sufficiently soluble at the concentration required except cystine and tyrosine. These will need initially to be dissolved in either 1 M NaOH (or 1 M HCl) to aid solution before adding to the stock solution (see Table 3.2(a)). It may be preferred to make up all the poorly soluble amino acids and the cystine and tyrosine as a separate solution in either 1.0 M HCl or 1 M NaOH. Add the amino acids one at a time to 75% of the final volume of solvent, preheated to 50 °C, to aid solution. When all have dissolved, allow the solution to cool to room temperature and adjust the pH to 10.0 with 1 M NaOH. Make up to the final volume and aliquot in appropriate amounts relative to the usual batch size of medium. If stored frozen at −20 °C or below, there is no need to filter sterilize. Storage above −20 °C is not recommended for any length of time but if short-term storage is to be at 4 °C the solution must be filter sterilized.

Balanced salt solution (see Tables 3.3 and 3.4)

Prepare the appropriate stock salt solutions for the type of medium required. All the salts are soluble in purified water. To make each stock solution add each salt in turn to 70% of the final volume of water and when dissolved make up to the final volume. Store at 4 °C. The basal medium Eagle (BME) and minimal essential medium (MEM) use Earle's salts.

Glutamine, 100 × (200 mM)

L-glutamine 29.22 gr
Purified water to 1 litre

Filter sterilize and aliquot aseptically. Store at −20 °C.
The method of preparation chosen will depend on individual requirements. See Tables 3.2(b). and 3.2(c). for levels of use.
(Glutamine is unstable above freezing, where it undergoes spontaneous breakdown in the absence of cells to form pyrollidone carboxyllic acid. The speed of the reaction is pH and temperature dependent. The level of available glutamine will drop to less than 50% after 5 days at 37 °C.)

Table 3.2. *Amino acids: solubilities and stock solutions*

(a) 100× Solutions

Amino acid	FW	Maximum solubility (at 20°C) (g/litre)	Solvent	BME	MEM	DMEM	RPMI	McCoy	Hams
L-alanine	89.09	157.80	Water	–	–	–	–	1.336	0.891
L-arginine·HCl	210.70	156.70	Water	2.106	12.64	8.40	20.0	4.214	21.07
L-asparagine	132.10	23.50	Water	–	–	–	5.682	4.503	1.30
L-aspartic acid	133.10	4.18	Water^a	–	–	–	2.00	1.997	1.33
L-cysteine	121.20	121.20	Water	–	–	–	–	2.424	3.60
L-cystine	240.30	120.0	1 M HCl★	1.201	2.402	4.80	5.00	–	–
Glutamic acid	147.10	7.40	Water^a	–	–	–	2.00	2.207	1.471
Glycine	75.07	225.2	Water^a	–	–	3.00	1.00	0.751	0.751
L-histidine	155.20	42.90	Water^a	0.80	4.19	4.20	1.50	2.096	2.096
L-hydroxyproline	131.10	345.2	Water	–	–	–	2.00	1.967	–
L-isoleucine	131.20	33.60	Water	2.623	5.25	10.48	5.00	3.936	0.394
L-leucine	131.20	23.40	Water^a	2.623	5.25	10.48	5.00	3.936	1.312
L-lysine·HCl	182.70	666.0	Water	3.653	7.306	14.62	4.00	3.654	3.654
L-methionine	149.20	51.40	Water^a	0.746	1.49	3.00	1.50	1.492	0.448
L-phenylalanine	165.20	27.35	Water	1.651	3.302	6.60	1.50	1.652	0.496
L-proline	115.10	1545.0	Water	–	–	–	2.00	1.727	3.453
L-serine	105.10	362.0	Water	–	–	4.20	3.00	2.628	1.051
L-threonine	119.10	16.30	Water	2.382	4.764	9.52	2.00	1.787	1.191
L-tryptophan	204.20	10.57	Water^a	0.408	1.02	1.60	0.50	0.306	0.204
L-tyrosine	181.20	91.00	0.5 M HC^b	1.811	3.622	7.20	2.00	1.812	0.544
L-valine	117.10	56.50	Water^a	2.343	4.690	9.36	2.00	1.757	1.171

Notes:

[a] Although soluble in water at the concentrations required, the addition of a small amount of 0.5 M HCl or NaOH may aid solution. It may be preferred to make up these amino acids as a separate solution.

[b] Or 0.5 M NaOH

Table 3.2. (cont.)

(b) 50X solution

| L-glutamine | 146.10 | 36.00 | Water | 14.615 | 14.615 | 29.20 | 15.00 | 10.958 | 7.31 |

(c) Glutamine 100X solution (200 mM) (29.22 g/l)

BME (mg/ml)	(ml/l)	MEM (mg/ml)	(ml/l)	DMEM (mg/ml)	(ml/l)	RPMI (mg/ml)	(ml/l)	McCoys 5A (mg/ml)	(ml/l)	Hams F12 (mg/ml)	(ml/l)
292.30	1.00	292.30	1.00	584.00	2.00	300.00	1.02	219.50	0.75	146.20	0.50

Table 3.3. *Individual stock salt solutions for media production*

Chemical	%	X conc.	BME[1] (ml/l)	MEM[2] (ml/l)	DMEM[3] (ml/l)	RPMI[4] (ml/l)	McCoys 5A[5] (ml/l)	Hams F.12[6] (ml/l)
KCl	20.0	500.0	2.0	2.0	2.0	2.0	–	1.12
CaCl2	20.0	1000.0	1.0	1.0	1.0	–	0.5	0.166
$Ca(NO_3)_2$	10.0	1000.0	–	–	–	1.0	–	–
$Fe(NO_3)_3 \cdot 9H_2O$	0.01	10.0	–	–	100.0	–	–	–
$FeSO_4 \cdot 7H_2O$	0.834	10.0	–	–	–	–	–	100.0
$MgCl_2$ anhydrous	2.861	500.0	–	–	–	–	–	2.0
$MgSO_4 \cdot 7H_2O$	20.0	1000.0	1.0	1.0	1.0	0.5	1.0	–
$NaHCO_3$	22.0	100.0	10.0	10.0	16.80	9.01	10.0	5.35
$NaH_2PO_4 \cdot H_2O$	14.0	1000.0	1.0	1.0	–	–	–	–
$NaH_2PO_4 \cdot H_2O$	12.5	1000.0	–	–	1.0	–	–	–
$NaH_2PO_4 \cdot H_2O$	29.0	500.0	–	–	–	–	2.0	–
$ZnSO_4 \cdot 7H_2O$	0.863	10.00	–	–	–	–	–	100.0

Sources:

[1] Eagle (1955).

[2] Eagle (1959).

[3] Dulbecco & Freeman (1959).

[4] Moore, Gerner & Franklin (1967).

[5] McCoy, Maxwell & Kruse (1959).

[6] Ham (1965).

Table 3.4. *Balanced salt solutions*

Component	Earle's[1] EBSS (mg/l)	Hanks'[2] HBSS (mg/l)	Gey's[3] GBSS (mg/l)	Dulbecco's[4] PBS (mg/l)	Puck's A[5] Saline (mg/l)	Puck's G[6] Saline (mg/l)
$CaCl_2·2H_2O$	264.90	185.50	225.00	132.50[b]	–	–
KCl	400.00	400.00	375.00	200.00	400.00	400.00
KH_2PO_4	–	60.00	–	200.00	–	–
$MgCl_2·6H_2O$	–	–	210.00	100.00[c]	–	–
$MgSO_4·7H_2O$	200.00	200.00	70.00	–	–	–
NaCl	6800.00	8000.00	7000.00	8000.00	8000.00	8000.00
$NaHCO_3$	2200.00	350.00	2270.00	–	350.00	–
Na_2HPO_4	–	47.68	120.00	1150.00	–	–
$NaH_2PO_4·H_2O$	140.00	–	30.00	–	–	–
$Na_2HPO_4·7H_2O$	–	–	–	–	–	154.00
D-Glucose	1000.00	1000.00	1000.00	–	1000.00	1000.00
Phenol Red	10.00	10.00	10.00	–	5.00	5.00

Notes:
[b] The $CaCl_2$ is prepared separately=PBS 'B'.
[c] The $MgCl_2$ is prepared separately=PBS 'C'.
Sources:
[1] Earle *et al* (1943).
[2] Hanks & Wallace (1949).
[3] Gey, Coffman & Kubicek (1952).
[4] Dulbecco & Vogt (1954).
   The formula as given is for the complete BSS. Usually prepared as three solutions: PBS 'B'($CaCl_2$), and PBS 'C' ($MgCl_2$), and PBS 'A' (the remainder of the salts).
[5] Puck, Cieciura & Fischer (1957).
[6] Puck, Cieciura & Robinson (1958).

Penicillin/streptomycin solution, 100 ×

Penicillin G $\quad\quad$ $1.0 × 10^6$ units
Streptomycin $\quad\quad$ 1 g
PBS 'A' or 0.9% saline to 100 ml

Sterilize by filtration. Add 10 ml per litre to sterile medium.

Other antibiotics

There are a large number of antibiotics available on the market that are suitable for use with cell cultures. Table 3.5 lists a selection of the most frequently used ones with information on their preparation and use.

Table 3.5. *Antibiotics used in cell culture*

Antibiotic	Molecular weight	Solvent	Storage	Stock soln.	Stock (grams/100 ml)	Cytotoxic concentration[1] (μg/ml)	Use concentration (μg/ml)	Estimated effective life in media at 37°C (days)	Gram-positive	Gram-negative	Myco-plasma	Moulds/yeasts
Penicillin G, sodium salt	356.4	a, b, c	0°C	100X	10⁴ U	10000 U	100 U.	3★	X			
Streptomycin SO$_4$	1457.4	a, b, c	0°C	100X	1.00	30000	100	4★	X	X		
Dihydrostreptomycin SO$_4$	730.7	a, b, c	0°C	100X	1.00	30000	100	5	X	X		
Ampicillin, sodium salt	371.4	a, b, c	2-8°C	100X	1.00		100	3	X	X		
Gentamycin SO$_4$	543.0	a, b, c	2-8°C	200X	1.00	3000	50-100	5	X	X		
Kanamycin SO$_4$	582.6	a, b, c	2-8°C	100X	1.00	10000	100	5	X	X	X	
Neomycin SO$_4$	908.9	a, b, c	2-8°C	200X	1.00	3000	50	5	X	X		
Chloramphenicol[2]	323.1	d	RT	100X	0.05	30	5	8	X	X		
Chlortetracycline·HCl	515.3	a, b, c	0°C	50X	0.5	80	10	2	X	X	X	
Hydroxytetracycline·HCl	480.9	a, b, c	0°C	100X	0.1	25	5-10	4	X	X	X	
Polymixin B SO$_4$	1202.5	a, b, c	2-8°C	33X	10 000 U	3000	50(300 U)	5	X	X		
Tylosin tartrate	1066.2	a, b, c	2-8°C	1000X	0.8	300	8-10	3	X		X	
Fungizone[3][4]	924.11	a, b, c	2-8°C	100X	0.025	30	2.5	3				X
Amphotericin B	924.11	DMSO	2-8°C	100X	0.025	30	2.5	3			X	X
Nystatin	926.1	a, b, c	0°C	100X	10⁶ U	600	50(1000 U)	3				X

Notes:

[1] May vary from cell line to cell line. [2] Use only if no other antibiotic is effective. [3] Trade mark of E.R. Squibb & Co [4] requires 20.5 mg/100 ml sodium deoxycholate in the stock solution.

Solvents: [a] PBS 'A' [b] 0.9% Saline [c] purified water [d] 0.5 M NaOH

★ Sigma have recently brought out a stabilized Penicillin/Streptomycin solution that will store for up to 40 days at 4°C. with minimal loss of activity.

X Denotes effective against organism.

Phenol Red (phenolsulphonphthalein), 1% (Indicator: pH range 6.8–8.4; yellow to red to purple)

Dissolve 1 g of the indicator in 32.5 ml of 0.1 M NaOH and stir for 15 min to dissolve.

Make up to 100 ml with purified water. Filter through paper and aliquot.

Not included in some media where the colour may interfere with other aspects of a study. A common contaminant of Phenol Red exerts an oestrogenic effect which may interfere with hormonal studies.

Vitamins, 100 ×

All of the vitamins required except folic acid are readily soluble in water (see Table 3.6). Add the vitamins one at a time to 75% of the final volume of purified water preheated to 50 °C to aid solution. When all are dissolved, add the folic acid prepared separately in 0.2 M NaOH. Make up to the final volume, mix well, and aliquot in appropriate amounts relative to the usual batch size of medium. If stored frozen at −20 °C or below, there is no need to filter sterilize. Storage above −20 °C is not recommended for any length of time but, if short-term storage is to be at 4 °C, the solution must be filter sterilized.

Batch preparation

It is recommended that a checklist is always used when preparing medium from basic ingredients to minimize the opportunity for mistakes to occur. Each ingredient should be checked off as it is added, thereby reducing the chance of omissions or double additions occurring.

1 If using frozen aliquots of some of the ingredients, remove sufficient of each of the ingredients from −20 °C storage and thaw them in a water bath set at 37 °C (see comments regarding water baths in Chapter 2). When fully dissolved, mix well.
2 To approximately 75% of the final volume of purified water, add, in order, the correct amount of the stock solutions: salts, amino acids, vitamins, glutamine, Phenol Red.
3 Weigh out the remaining salts and the remainder of the other ingredients. Add to the water and stir until they are dissolved.
4 If using NaOH or HCl to adjust the pH, do so at this stage.
5 Make up the volume to 100%.
6 If adjusting the pH using carbon dioxide gas, do so at this stage.

Table 3.6. *Vitamins: solubilities and 100× stock solutions*[1]

Vitamin	Formula weight	Solvent	Maximum solubility (g/l)	BME	MEM	DMEM	RPMI 1640	McCoys 5A	Hams F12
Ascorbic acid (Vitamin C)	176.10	Water	330.0	–	–	–	–	56.0	–
Biotin (Vitamin H)	244.31	Water	0.2	100.0	–	–	20.0	20.0	0.73
D–Ca pantothenate	476.55	Water	350.0	100.0	100.0	400.0	25.0	20.0	48.0
Choline chloride	139.60	Water	>50.0	100.0	100.0	400.0	300.0	1000.0	1396.0
Folic acid (Vitamin M)	441.40	0.2 M NaOH	100.0	100.0	100.0	400.0	100.0	1000.0	132.4
i-inositol	180.16	Water	150.0	180.0	200.0	720.0	3500.0	3600.0	1802.0
Niacinamide (Nicotinamide)	122.13	Water	1000.0	100.0	100.0	400.0	100.0	50.0	3.70
Nicotinic acid (Niacin)	123.11	Water	16.0	–	–	–	–	50.0	–
p–Aminobenzoic acid	137.10	Water	5.00	–	–	–	100.0	100.0	–
Pyridoxal HCl	203.62	Water	50.0	100.0	100.0	400.0	100.0	50.0	–
Pyridoxine HCl (Vitamin B_6/Pyridoxol)	205.64	Water	220.0	–	–	–	–	50.0	6.20
Riboflavin (Vitamin B_2)	376.37	Water	0.13	10.0	10.0	40.0	20.0	20.0	3.76
Thiamine HCl (Vitamin B_1/Aneurine)	337.27	Water	1000.0	100.0	100.0	400.0	100.0	20.0	33.70
Vitamin B_{12} (Cyanobalamin)	1355.42	Methanol	12.50	–	–	–	0.50	200.0	135.7

[1] Figures quoted are in milligrams per litre.

The medium should be filtered and dispensed aseptically as soon as possible after preparation.

Sterilizing the medium

All the procedures must be performed aseptically and with care. The wearing of disposable gloves is recommended. Spraying the gloves when on the hand with 70% ethanol is an added precaution. The following are two examples of batch preparation:

Small batches

This example involves the use of vacuum. (If using a water vacuum pump, always fit a trap between the pump and the medium reservoir so that water does not get drawn back into the flask containing the medium.)

1 If not already on switch on, the laminar flow cabinet and allow to run for 10 minutes before use.
2 Place the sterilized filter flask and holder into the cabinet.
3 Carefully remove and discard the wrapping.
4 Connect the line from the side arm to the vacuum source (this may need support).
5 Pour medium into the filter funnel (reservoir).
6 Undo the clamp on the vacuum line and switch on the vacuum.
7 When all the medium has passed through the filter, clamp off the line and switch off the vacuum source.
8 Leave for a few minutes to allow the vacuum within the flask to equilibrate.
9 Carefully remove the silicone rubber bung containing the filter housing and put to one side.
10 Carefully and aseptically decant the medium into sterile bottles.
11 Tighten all the bottle caps.
12 Label each bottle clearly with the name of the medium, date of preparation and batch number.
13 Store at 4 °C whilst quality control tests are performed (see further).

Larger batches

This example uses a mixing tank, peristaltic pump and a reusable glass dispensing bell. For larger batches, the use of a metered dispenser programmed to deliver a set volume in conjunction with a hand or foot pedal is very useful but not essential. This method is for those who do not have this facility.

1 If not already on, switch on the laminar flow cabinet and allow to run for 10 minutes before use.
2 Place the sterilized filter housing(s) into the cabinet and attach to a retort stand (if appropriate).
3 Carefully remove the wrapping from around the dispensing bell and very carefully clamp the bell to a retort stand. Discard the wrapping.
4 Before connecting the inlet tube switch on the peristaltic pump and adjust the speed to deliver medium at a manageable rate. Switch off the pump.
5 Remove the aluminium foil cover from the inlet tube and connect to the medium tank via the peristaltic pump.
6 Place all the sterile bottles to one side of the filter housings. Loosen the caps so that they rest on the neck.
7 Start the pump and open the tap on the dispensing bell[1].
8 Fill the first bottle. Close the tap.
9 Place the next bottle under the bell and open the tap. Fill the bottle and close the tap. Repeat this process until all the medium has been dispensed.
10 When filling is complete, label all the bottles with the name of the medium, date of preparation and batch number.

After filling, take sample(s) for quality control testing (see further) and store the remainder at 4 °C until quality control is complete.
Before use the medium will usually be supplemented aseptically with serum and antibiotic.

Note

[1] With practice, it is very easy to develop a routine and a rhythm to the process. Here, the caps of the first and second bottles are removed and placed top down on a piece of tissue paper soaked in 70% ethanol at the start of the process. The level of the dispensing bell is adjusted so that it just clears the top of the bottles, and both bottles are placed next to each other under the bell with the first bottle immediately under the outlet. The pump is switched on and the medium allowed to fill the first bottle to the required level. At this point both bottles are moved sideways so that the medium flow continues into the second bottle without interruption. Whilst the second bottle is filling, the cap is replaced on to the first bottle and the bottle placed to one side. A third bottle is taken from the other side, the cap removed and placed on the tissue paper and the bottle placed beside the second

bottle. This process is repeated until all the medium has been dispensed. This method has been used in our laboratory for dispensing hundreds of litres of medium each week for several decades without any contamination problems.

Other solutions

Growing and subculturing mammalian cells will also require a selection of other solutions.

HEPES buffer, I M

(N-2-hydroxyethyl-piperazine-N^1-2-ethanesulphonic acid $pK_a=7.31$ at 37 °C)
HEPES 238.3 g
purified water to 1 litre
pH 7.2–7.4

NB The pK_a of HEPES varies inversely with temperature by 0.014/°C. To attain a pH of 7.3 at 37 °C, the pH should be adjusted to 7.5 at room temperature (23 °C). Sterilize by filtration.
Use at a level of 10–25 mM (10–25 ml/litre) (Cytotoxic concentration=100 mM). Hanks's BSS. requires 10 mM, Hams F.12 20 mM and Earles BSS 40 mM. A general satisfactory compromise is to use 20 mM in all media. It is recommended that, when using HEPES, the level of sodium bicarbonate in the medium should be reduced to a level of no more than 850 mg/litre. Media containing HEPES buffer equilibrate with air.

Nigrosine

Used as an exclusion dye for assessing the health and viability of cells in conjunction with a counting chamber. Healthy cells will exclude the dye: sick or dying cells will take up the dye.

Nigrosine (Acid Black 2) 1 g
PBS 'A' to 100 ml

Filter through paper and aliquot. Sterilize by autoclaving.

Trypan Blue, 0.4%

Used as an exclusion dye for assessing the health and viability of cells in conjunction with a counting chamber. Healthy cells will exclude the dye: sick or dying cells will take up the dye.

Trypan Blue	0.4 g
NaCl	0.81 g
K_2HPO_4	0.06 g
PBS 'A' to 100 ml	

Sterilize by filtration. Aliquot and store at 4 °C.

Trypsin (1 : 250) 0.25%

This is the most popular proteolytic enzyme used to separate anchorage dependent cells from the surface on which they are growing and from one another into a suspension suitable for counting and subculturing. Purified trypsin alone is usually ineffective for tissue dissociation. Commercial 'Trypsin', e.g. Difco 1:250, is in fact pancreatin, a crude mixture of proteases, polysaccharides, nucleases and lipases extracted from porcine pancreas. Trypsin 1:250, the most commonly used trypsin preparation, has a level of activity sufficient to bring about the proteolytic digestion of 250 times its own weight of casein under assay conditions specified by the US National Formulary.

Prolonged exposure to trypsin will damage the cell membrane. The cell separation process should be monitored closely and the proteolytic activity of the trypsin neutralized as soon as possible by resuspension of the cells in medium containing serum.

To prepare 1 litre:

1 Tris saline

NaCl	8.0 g
KCl	0.38 g
Na_2HPO_4	0.1 g
Dextrose	1.0 g
Trizma base	3.0 g
Phenol Red, 1%	1.5 ml.

Dissolve in 200 ml purified water. Add 1 M HCl (approx. 1.25 ml per litre) to bring the pH to 7.7 at room temperature. Add:

2 Penicillin G, Na salt	100 000 Units
3 Streptomycin	100 mg

4 To 200 ml of purified water add:

Trypsin (1:250)	2.5 gr

Stir gently to dissolve. Assist solution by bubbling air through the mixture. Add to the Tris saline. Make up the volume to 1 litre with purified water.

Recheck the pH to 7.7 and adjust if necessary. Prefilter through a series of clarifying filters 1.2, 0.8 and 0.4 μM. Sterilize by filtration through a 0.22 μM filter. Bottle aseptically into aliquots and store at -20 °C. Samples should be tested for sterility (see further) and for tryptic efficiency by incubating dilutions with gelatin charcoal discs at 37 °C. Since these are no longer available commercially, they have to be prepared in-house using the method of Kohn (1953). In those bottles where digestion occurs, the charcoal granules are released and can be swirled around making determination of the end point very easy. Some organizations test the activity of their in-house trypsin using individual drops from a Pasteur pipette on to the coated side of exposed black and white photographic film. Where digestion of the gelatin retaining the black-coloured coat of silver salt occurs, the film dissolves away to reveal white plastic again making the assessment of the end point very easy.

Collagenase (Clostridiopeptidase A), chymotrypsin, elastase and hyaluronidase have also been used for the dissociation of tissues. A number of other proteases used in the subculture of primary cells are to be found in individual volumes in this series.

Trypsin/versene

To 16 ml of Versene, add 4 ml of trypsin 1:250.
NB Commercial trypsin/versene contains 0.5 g of trypsin and 0.2 g EDTA per litre in Hanks's BSS.

Versene (EDTA) 1:5000 (0.526 mM)

Chelating agent used to interact with the ionic bonds between anchorage-dependent cells and the substrate on which they are growing.

EDTA (versene), tetrasodium salt	200 mg
Phenol Red	15.0 mg
PBS 'A' to 1 litre.	

Check the pH to 7.2. Bottle in 16 ml aliquots in Universal bottles and sterilize by autoclaving at 121 °C for 15 minutes. Store at 4 °C.

Quality control

Liquid media from commercial suppliers comes ready for use with full quality control but, where medium is produced in-house from a commercial powder or raw ingredients, the following quality control tests should be conducted prior to use. The process of quality control starts at the beginning of the planning stage and continues for some time after production. The main points to be considered are listed below.

Chemicals

The choice of supplier and the quality of the chemicals used are very important. A recognized supplier of laboratory chemicals offering a range of products of adequate purity should be chosen. Some companies supply tissue culture-tested chemicals, e.g. Sigma. It is important to remember that some chemicals have limited shelf-lives, and consideration should be given to purchasing these in amounts that will normally be consumed within the stated shelf-life in order to avoid waste or mistakes. Amino acids and vitamins can be made up as 100 × stock solutions and stored frozen in aliquots to avoid wastage.

Water

The water used to prepare cell culture media must be either double distilled or deionized and of adequate purity. The main test for purity is the measurement of resistivity/conductivity. Water with a minimum resistance of 1.0 megohm/cm (1.0 µS/cm conductivity) (equivalent to double distilled water) or greater has in our experience proved satisfactory. Modern deionizers regularly produce water up to 18 megohm/cm quality, although there are some who consider that it is not necessarily an advantage to have such pure water for the preparation of cell culture media.

Filtration equipment

When filters are steam sterilized in their housings, there is always the possibility damage may occur to the membrane that might not be detectable to the eye. When using the smaller assemblies, it is not possible to check the integrity of the filter membrane without compromising sterility. In larger-scale production, where cartridge filters are used in series, usually as a pair, the integrity of the assembly can be tested without compromising the sterility. Some manufacturers supply equipment for integrity testing their cartridge filters after sterilization but before use. See Chapter 2 for a list of the main suppliers of filtration equipment.

Sterility testing

Before using any batch of medium produced in-house, it is vital to ensure that the process has produced medium that is sterile. For in-house production it is not necessary to comply fully with the British, European or US Regulations for the sterility testing of pharmaceutical products. For larger batches, a bottle of filled medium should be taken from the start, middle and end of the filling process. For smaller batches, only the end product need be tested. At each stage inoculate 20 ml samples aseptically into 200 ml bottles containing 80 ml of test culture medium. Contaminating micro-organisms fall into one of four main categories:

1 aerobic bacteria: require an atmosphere containing normal levels of oxygen to grow;
2 facultative anaerobes: will grow in an atmosphere containing reduced levels of oxygen;
3 anaerobic bacteria: will only grow in an atmosphere with very low levels of oxygen;
4 yeasts and moulds (fungi).

To test for these organisms, only two types of medium are required:

1 A fluid mercaptoacetate medium (FMM) for anaerobic (and facultative anaerobic) bacteria. One bottle of FMM is required per test sample incubated at 30–35 °C.
2 Soya bean casein digest medium (SBC) for aerobic bacteria and yeasts and moulds. Two bottles of SBC are required per test sample one incubated at 20–25 °C (yeasts/moulds) and the other at 30–35 °C. (aerobic bacteria).

Controls

Ideally, the following controls should be set up alongside the samples under test:

1 Samples of each batch of test microbiological medium should be inoculated with positive cultures to confirm they are able to support the growth of the appropriate organisms:
FMM: anaerobic and facultative anaerobic organisms;
SBC: aerobic bacteria – incubated at 35 °C; yeast and/or fungi – incubated at 25 °C.
2 as 1 above but also containing 20 ml samples of the medium.

All bottles should be incubated for a minimum of 7 days but preferably 14 days. They should be examined daily for any signs of growth.

pH

In most mammalian species the normal *in vivo* pH of the cell and its environment is 7.2-7.4 and, as a result, most cell culture media are designed to have a pH within the range pH 7.0–7.5. The pH of every batch of medium must be checked and adjusted preferably using a reliable and accurate pH meter, ideally one with automatic temperature compensation.

The inclusion of Phenol Red as indicator in the majority of cell culture media enables the approximate pH of any bottle of medium to be assessed quickly by visual inspection after sterile filling. This indicator spans the pH range from 6.8 (yellow) to 8.4 (purple). At pH 7.2-7.4, it is a deep red colour. Until familiarity with the colour/pH relationship of the dye has been achieved, test bottles can be tested with a pH meter or by comparison with the standards in a comparator.

Osmolarity testing

Cell culture media for mammalian cells are not only formulated to provide adequate levels of nutrients and ions required for normal cell metabolism at the correct pH but the concentrations of the ingredients, particularly the salts in the balanced salt solution are designed to provide an osmotic pressure similar to that found *in vivo*. This is measured in milli-osmoles (mOsm). Apparatus is available for testing the osmolarity of aqueous solutions such as cell culture media. Each medium has a standard osmolarity with upper and lower limits, which will vary slightly from supplier to supplier and batch to batch. Details of expected values are given in Table 3.7. Media with an osmolarity on the lower side is much less damaging to the cells than one that is on the higher side. Over 400 mOsm, plating efficiency will drop off markedly. If an osmometer is not available, generally, if the medium has been made up carefully and accurately, it will be within the accepted limits.

Plating efficiency

This is dealt with in Chapter 4.

Table 3.7. *Osmolarity values for cell culture media and BSS*★

BME (Earles salts)	282–318	McCoys 5A	275–320
BME (Hanks's salts)	270–310	Hams F12	285–332
MEM (Earles salts)	278–314	Earles BSS	271–308
MEM (Hanks's salts)	274–317	Hanks's BSS	265–305
DMEM	322–374	Geys BSS	265–305
RPMI 1640	268–319	Dulbecco's PBS (complete)	280–311
		(Human whole blood	285–308)

Notes:
Above values are quoted in milli-osmoles (mOsm) per litre.
★ Values quoted by commercial suppliers and others tend to vary a small amount according to the source.
Two measures exist for osmotically active particles (OAPs) in an aqueous solution:

Osmolarity: relates to the volume of solvent (mOsm per litre of water). One osmole is the mass of 6.023×10^{23} OAPs in an aqueous solution.
Osmolality: relates to the weight of solvent (mOsm per kg of water). One milli-osmole per kg of water (mOsm/kg water) produces a freezing point depression of 0.001858°C; at 38°C. This is equivalent to a pressure of 19 mm of mercury.

In practice, there is very little difference between osmolarity and osmolality when water is the solvent.
Sources:
Ogston & Wells (1970).
Preston, Davis & Ogston (1965).
Waymouth (1970).
Waymouth (1973).
Whitehead (1976).

Suppliers of comparators

The Tintometer Ltd, Waterloo Road, Salisbury, Wiltshire SP1 2JY, UK   Tel: 01722 327242   Fax: 01722 412322

Suppliers of cell culture media
Biowhittaker Inc., 8830 Biggs Ford Road, Walkersville, Maryland 21793, USA Tel: (301)-898-7025   Fax: (301)-845-8338   International Fax: (301)-845-8291
BioWhittaker UK Ltd, BioWhittaker House, 1 Asheville Way, Wokingham, Berkshire RG41 2PL, UK   Tel: 0118 9795234   Fax: 0118 9795231
Harlan Sera-Lab Ltd, Hophurst Lane, Crawley Down, East Sussex RH10 4FF, UK Tel: 01342 716366   Fax: 01342 717351

Hy-Clone Laboratories, International Dept, 1725 South Hyclone Road, Logan, Utah 84321, USA   Tel: 1-801-753-4584   Fax: 1-801-753 4589/1-801-750–0809

HyClone Europe, Dendermondsesteenweg 56, 9300–Aalst, Belgium Tel: +32.53.706090   Fax: +32.53.704171

ICN Pharmaceuticals Inc., 3300 Hyland Avenue, Costa Mesa, California 92626, USA   Tel: (714)-545-0113   Fax: (800)-334-6999

ICN Biomedicals Ltd, Unit 18, Thame Park Business Centre, Wenman Road, Thame, Oxfordshire OX9 3XA, UK   Tel: 01844 213366   Fax: 01844 223399

Imperial Laboratories (Europe) Ltd, West Portway, Andover, Hampshire SP10 3LF, UK   Tel: 01264 333311   Fax: 01264 332412

Irvine Scientific, 2511 Daimler Street, Santa Ana, California 92705-5588. USA   Tel: (714)-261-7800   Fax: (714)-261-6522

Life Technologies Inc., 8451 Helgerman Court, Gaithersburg, Maryland 20877-9980, USA   Tel: (301)-840–8000   Fax: (301)-258-8238

Life Technologies Ltd, 3 Fountain Drive, Inchinnan Business Park, Paisley PA4 9RF, Scotland, UK   Tel: 0141-814 6100   Fax: 0141-814 6258/6260

PAA Biologics Ltd, 9 Derwentside Business Centre, Villa Real, Consett, Co. Durham DH8 6BP, UK   Tel: 01207 582993   Fax: 01207 583002

Sigma-Aldrich Company Ltd, Fancy Road, Poole, Dorset BH12 4QH, UK   Tel: 01202 733114   Fax: 01202 715460

Sigma Chemical Company, PO Box 14508, St.Louis, Missouri 63178, USA   Tel: (314)-771-5750   Fax: (314)-771-5757

TCS Biologicals Ltd, Botolph Claydon, Buckingham MK18 2LR, UK   Tel: 01296 714071   Fax: 01296 714806

Suppliers of laboratory grade chemicals

Aldrich Chemical Company, 940 West St Paul Avenue, Milwaukee, Wisconsin 53233, USA   Tel: (414)-273-3850   Fax: (414)-273-2094

BDH=Merck Ltd, Merck House, Poole, Dorset BH15 1TD, UK   Tel: 01202 669700   Fax: 01202 666541

Fisher Chemicals, Bishop Meadow Road, Loughborough, Leicestershire LE11 ORG, UK   Tel: 01509 231166   Fax: 01509 231893

Fluka Chemicals, The Old Brickyard, New Road, Gillingham, Dorset SP8 4JL, UK   Tel: 01747 823097   Fax: 01747 824596

Fluka Chemie AG, Industriestrasse 25, CH-9471, Buchs, Switzerland   Tel: +41/81 755 2511   Fax: +41/81 756 5449

GFS Chemicals, PO Box 245, Powell, Ohio 43065, USA   Tel: (614)-881-5501   Fax: (614)-881-5989

ICN Biomedicals Ltd, Unit 18, Thame Park Business Centre, Wenman Road, Thame, Oxfordshire OX9 3XA, UK   Tel: 01844 213366   Fax: 01844 213399

ICN Pharmaceuticals Inc., 3300 Hyland Avenue, Costa Mesa, California 92626, USA   Tel: (800)-854-0530   Fax: (800)-334-6999

E.Merck KgaA., PO Box 4119, 64271-Darmstadt, Frankfurter Strasse 250, Germany   Tel: +49.61.51.720   Fax: +49.61.51.72.2000

Sigma Chemical Company, PO Box 14508, St.Louis, Missouri, 63178, USA   Tel: (314)-771-5750   Fax: (314)-771-5757

Sigma-Aldrich Company Ltd, Fancy Road, Poole, Dorset BH12 4QH, UK   Tel: 01202 733114   Fax: 01202 715460

References

Dulbecco, R. & Vogt, M. (1954). Plaque formation and the isolation of pure cell lines with poliomyelitis viruses. *J. Exp. Med.*, **99**, 167–82.

Dulbecco, R. & Freeman, G. (1959). Plaque production by the Polyoma virus. *Virology*, **8**, 396–7.

Eagle, H. (1955). Nutritional needs of mammalian cells in culture. *Science*, **122**, 501.

Eagle, H. (1956a). Myo-inositol as an essential growth factor for normal and malignant human cells in tissue culture. *J. Biol. Chem.*, **214**, 845–7.

Eagle, H. (1956b). Propagation in a fluid medium of a human epidermoid carcinoma, Strain KB. *Proc. Soc. Exp. Biol. Med.*, **89**, 362–4.

Eagle, H. (1959). Amino acid metabolism in mammalian cell cultures. *Science*, **130**, 432–7.

Earle, W. R. (1943). Production of malignancy in-vitro. IV. The Mouse fibroblast cultures and changes seen in the living cells. *J. Natl Cancer Inst.*, **4**, 165–212.

Gey, G. O., Coffman, W. D. & Kubicek, M. T. (1952). Tissue culture studies of the profilerative capacity of cervical carcinoma and normal epithelium. *Cancer Res.*, **12**, 364–5.

Good, N. E/, Winget, G. D., Winter, W., Connolly, T. N., Izawa, S. & Singh, R. M. M. (1966). Hydrogen ion buffers and biological research. *Biochemistry*, **5**, 467–77.

Ham, R. G. (1965). Clonal growth of mammalian cells in a chemically defined synthetic medium. *Proc. Natl Acad. Sci. USA*, **53**, 288–93.

Hanks, J. H. & Wallace, R. E. (1949). Relation of oxygen and temperature in the preservation of tissues by refrigeration. *Proc. Soc. Exp. Biol. Med.*, **71**, 196–200.

Kohn, J. (1953). A preliminary report of a new gelatin liquefaction method. *J.Clin.Path.*, **6**, 249.

McCoy, T. A., Maxwell, M. & Kruse, P. F. (1959). Amino acid requirement of the Novioff hepatoma *in vitro*. *Proc. Soc. Exp. Biol. Med.*, **100**, 115-18.

Moore, G. E., Gerner, R. E. & Franklin, H. A. (1967). Culture of normal human leukocytes. *JAMA*, **199**, 519–24.

Neuman, R. E. & McCoy, T. A. (1958). Growth-promoting properties of pyruvate, oxalacetate and α-ketoglutarate for isolated Walker Carcinosarcoma 256 cells. *Proc. Soc. Exp. Biol-Med.*, **98**, 303–6.

Preston, S. N., Davis, M. & Ogston, A. G. (1965). The composition and physicochemical properties of hyaluronic acids prepared from ox synovial fluid and from a case of mesothelioma. *Biochem. J.*, **96**, 449–74.

Puck, T. T., Cieciura, S. J. & Fischer, H. W. (1957). Clonal growth *in-vitro* of human cells with fibroblastic morphology. *J. Exp. Med.*, **106**, 145–57.

Puck, T. T., Cieciura, S. J. & Robinson, A. (1958). Genetics of somatic mammalian cells. *J. Exp. Med.*, **108**, 945-54.

Ogston, A. G. & Wells, J. D. (1970). Osmometry with single Sephadex beads. *Biochem. J.*, **119**, 67–73.

Smith, J. D., Freeman, G., Vogt, M. & Dulbecco, R. (1960). The nucleic acid of polyoma virus. *Virology*, **12**, 185–96.

Waymouth, C. (1970). Osmolality of mammalian blood and of media for culture of mammalian cells. *In-Vitro*, **6** (2), 109–27.

Waymouth, C. (1973). Determination and survey of osmolality in cell culture media. In *Tissue Culture: Methods and Applications*. Ed. P. F. Kruse, Jr & M. K. Patterson, Jr, Chap. 6, pp. 703–9. New York and London: Academic Press.

Whitehead, R. H. (1976). Implications of the osmolalities of some commonly used tissue culture media. *Br. J. Cancer*, **33** (3), 347–8 (Letter).

4

Serum

Since the early days of cell culture technology, serum has been used as an important component in media for growing animal cells. It remains the principal supplement to increase the effectiveness of chemically defined media and contains most, if not all, of the growth factors and hormones that cells require for their growth. It is wide practice to use varying amounts of serum (up to 20%) to make up for small qualitative deficiencies in synthetic media.

Serum-free medium

In the early 1970s it was suggested that serum proteins acted mostly as carriers for low molecular weight hormones or growth factors and, if these could be provided in the medium, the presence of serum would not be necessary. In recent years many excellent serum-free media have been brought on to the market, which have proved this hypothesis. However, these have two disadvantages in the average laboratory set-up: *(a)* expense and *(b)* it is apparent that each cell line may require a different set of growth factors and base medium for growth. In a situation where one cell line is being grown on a large scale and deficiencies of particular components may be monitored on a regular basis and compensated for, serum-free media are useful, but where the researcher may be growing different cell lines the convenience of using serum cannot be overestimated.

Most serum-free media may be purchased as a complete package (either in liquid or powder form) and will contain a variety of amino acids, vitamins and inorganic salts plus one or more supplements to replace serum, e.g. bovine serum albumin, transferrin, insulin, epidermal growth factor, bovine pituitary extract, etc. ICN Flow laboratories have produced two products which are designed to replace serum and these may be added to any basal medium. The products, TCH™ (for use with human cells) and TCM™ (for use with other species including mouse, rat, hamster, bovine, feline and canine cells have a low protein content and contain no growth factors or steroid hormones. These serum-free systems would be useful where purification processes are required (because the low protein content simplifies downstream processing) or in a situation where a study involves growth factors or steroid hormones (ICN Flow sales leaflet).

Specialized serum-free media are marketed by Gibco BRL, e.g. serum free medium for the culture of human umbilical vein endothelial cells (HUVECS) (Battista, Bowen & Gorfien, 1995). Another new product from the same company is a serum-free medium for growth of BT1-TN-5B1-4 (cabbage looper *Trichoplusia* NI)cells (High Five cells). The traditional cell lines for the Baculovirus expression vector system have been SF9 and SF21 insect cells derived from the fall armyworm. Recently, recombinant protein expression levels in High Five cells has been reported to be up to ten-fold greater and, although current serum-free media formulations support suspension growth of High Five cells, they have different nutritional requirements from the classic cell lines and this new serum free medium has been optimized to achieve rapid doubling times and high peak cell densities resulting in higher recombinant protein yields (Godwin, Danner & Gorfien, 1995).

Since serum is the most expensive component of tissue culture careful consideration needs to be given to the selection of the best system for the work to be done, e.g. if one is buying serum, the cost of batch testing needs to be considered in comparison with the cost of buying serum free medium and any necessary additives. Although most companies will provide free samples for batch testing, it is necessary to consider the labour costs. The advantages of a lack of biological variability in serum-free media may preclude the need for batch testing but could work out as expensive as the purchase of serum, if the user is growing a wide variety of cells needing different additives and constant monitoring of separate components.

Types of serum

The most common types of sera used in cell culture includes foetal bovine serum (also known as foetal calf serum), newborn calf serum, donor calf serum, horse serum, chicken serum and human serum.

Foetal bovine serum

Foetal bovine serum is probably the most widely used of the above and will promote growth in a wide range of cell lines. It is a by-product of the meat industry and is only available from those countries where cattle are free to roam and the cows free to become pregnant.

Newborn calf serum

Newborn calf serum is processed from animals that are under 10 days of age. This serum is less expensive and has proved an efficient growth promoter in cells such as 3T3 J2 (mouse fibroblasts) (Todaro & Green, 1963).

Donor calf serum

Donor calf serum is obtained from processed whole blood of calves that are up to 8 months old, the animals being partially bled from the jugular vein on a routine basis. A number of cells with rapid doubling times will grow well in media supplemented with donor calf serum and, from an economic point of view, it is worth trying to wean cells from foetal bovine to donor serum if they show rapid growth characteristics. It should be noted however, that growing large volumes of suspension cells in donor calf serum can be disadvantageous as the increased protein content of the serum in large volumes causes the suspension to become somewhat opaque and renders counting cells and morphological observations difficult.

Other sera

Horse serum is used widely in combination with foetal bovine serum for the growth of PC12 cells (rat adrenal pheochromocytoma) and is used extensively in media for the testing of samples for the presence of mycoplasma. Human serum and chicken serum have specialized uses for those cells which require a serum derived from a similar species of animal, e.g. chicken serum is used to promote growth in a wide range of avian cells.

Sources

In the early 1990s the foetal bovine serum industry suffered a crisis of confidence. The progressive demand for foetal bovine serum (FBS) from the biopharmaceutical industry and a supply threatened by the incidence of diseases like BSE (bovine spongiform encephalopathy) created a motive for malpractice, therefore anyone buying FBS should care about its source. In 1991, the FDA (Food and Drug Administration) recalled two lots of FBS processed in the United States. The serum had arrived from Brazil labelled as human serum and thereby evading the USDA (United States Department of Agriculture) (Hodgson, 1991). The geographical source from which bovine serum originates is a critical consideration when purchasing serum. The source of serum used may now determine whether or not a biotherapeutic agent produced will receive approval by certain regulatory bodies including the FDA. In the USA, the USDA recognizes only five areas in the world from which foetal bovine serum imported into the USA may originate. It is worth considering that, if one is likely to need to send a cell line to the USA, there may be a problem if those cells have been grown in serum which is not approved by the USDA. The FDA specifically prohibits the use of FBS from countries like the UK where BSE is present and all the national medicines and veterinary control agencies in Europe have acted likewise, so as a general rule it is probably better to avoid purchasing UK-produced serum. As the rules set out by the governing bodies are subject to change, if it is known that a cell line is ever going to find its way to the USA, it is wise to check on the serum regulations before purchasing a batch for a particular use.

Once the country of origin has been decided upon, price and quality are the next most important considerations. Where price is concerned, it should be borne in mind that one is better able to negotiate a price if a large batch of serum is purchased and this is also beneficial from the point of continuity. The major effect of the animal and human healthcare regulations is to skew demand for serum toward the permissible sources and thereby to create large price differentials. At the top of the range is New Zealand, Australian and US serum, slightly cheaper than that is USDA approved serum (from Central America approved after quarantine) and further down the scale is serum from Europe, whilst Brazilian serum is at the bottom. It may be useful to note that the USDA and FDA have been known to request that cells entering the USA must be accompanied by a certificate which states that they have a history of growth in approved sera (so that it will be of little use to passage cells only using approved sera just before cells are exported to the States)

If it is essential to use serum of US origin and the price is prohibitive it may be worth looking at a mixture of FBS and newborn calf serum of US origin. In our laboratory, we have successfully grown a number of suspension cells, adhesion cells and hybridomas in 'Serum Supreme' a mixture as described above, marketed by Biowhittaker.

Batch testing

If there is no necessity for the more expensive sera, there are usually a variety of excellent European (non-UK) sera on the market at a reasonable price.

Whichever country of origin/suitably priced serum is chosen, it obviously must be one which is suitable for the desired purpose, which is why it is essential to batch test.

Most companies who supply serum will be happy to supply up to 500 ml of a serum for batch testing and, at the time of requesting a sample, it is sensible to ensure that batches will be accompanied by a statement giving the size of the batch, the country of origin and a certificate of analysis. If there is any likelihood of the latter not being produced, it is probably not worth including that particular sample in the batch testing. A typical certificate of analysis will give information on sterility testing for bacteria, yeasts and fungi, mycoplasma, bacteriophage and viruses such as Bovine viral diarrhoea, Parainfluenza 3 and infectious bovine rhinotracheitis. In addition to this, there should be a list of results from physical and biochemical analysis covering such items as pH at 37 °C, osmolarity, albumin, beta globulin, gamma globulin, haemoglobin, protein and endotoxin levels. If any commodity is known to be needed at a certain level, this information should be given to the supplier at the time of requesting samples. It may be helpful to note that, with the variation in batches, it is always advisable to approach more than one supplier even though the original supplier is of excellent reputation.

If serum is to be used for the growth of one particular type of cell, the requirements of that cell type must be considered (refer to appropriate book in the Handbooks in Practical Animal Cell Biology series.) However, if serum is needed for the growth of a variety of cells, there are a number of general tests which must be carried out on each sample and the best overall serum selected from the results. When selecting cells for batch testing a wide variety should be chosen, e.g. a suspension cell line which is a slow grower as well as one which has a rapid doubling time, a cell line which is known to be sensitive to any small change in growth conditions, a hybridoma, an epithelial cell line, a totally fibroblastic cell line, cells of different species, etc. Use a known serum as a control and set up all tests in at least triplicate flasks.

Tests for a general serum include growth/viability studies, cloning and plating efficiencies.

Suspension cell test

Preparation of cells

1 Count stock suspensions of each cell line to determine the number of viable cells per ml.

2 Prepare a dilution of each cell type in RPMI 1640+10% FBS to give a concentration of 2×10^5 cells/ml.

Preparation of test and control media

3 Using two bottles for each test and control medium dispense 8.5 ml RPMI and 1 ml foetal bovine serum into separate flasks (25 cm²)

The test

4. Add 0.5 ml cells at 2×10^5/ml to one pair of flasks containing test and control media.
Repeat this for each suspension cell line (final concentration will be 10^4 cells/ml in each flask).
5 Mix and incubate at 37 °C for 7 days in a CO_2 incubator.

Counting and calculating

6 Perform a viable count on each flask and determine the total number of viable cells in each. Record the results.
7 The test counts should be $>6 \times 10^5$/ml to be satisfactory.

Monolayer test

Preparation of cells (e.g. BHK Clone 13 and Swiss 3T3 cells)

1 Trypsinize the cells from monolayers. Resuspend in 10 ml DMEM (Dulbeccos' modified Eagles' medium)+10% FBS and perform a viable count.
2 Calculate the total number of viable cells and dilute to give 10 ml at 10^5/ml.
3 Add 1 ml of this suspension to 9 ml of complete DMEM to give 10 ml at 10^4/ml.

Preparation of test and control media

4 Using two bottles for each test and control medium, dispense 47.5 ml of medium followed by 2.5 ml of foetal bovine serum into separate bottles.(Note that this is 5% serum since the test is being carried out using established cell lines. It is necessary to use at least 10% serum for culturing most primary cells.)

The test

5 Add 0.25 ml of BHK Clone 13 cells at 10^4/ml to one bottle of each pair of test and control medium and 0.25 ml Swiss 3T3 cells at 10^4/ml to the other.

6 Mix well and dispense 10 ml into each of 4, 25 cm^2 flasks per sample.

7 Incubate in a CO_2 incubator at 37 °C for 7 days.

Staining

8 Decant the medium from each bottle and invert to drain with the cell sheet uppermost so the liquid drains down the opposite side to the cell sheet. Leave for 5 minutes.

9 Add 3 ml of Giemsa stain (see Chapter 8) to each flask and lay flat to allow the stain to come into contact with the cell sheet. Leave for 15–20 minutes.

10 Pour off stain, rinse with tap water and invert to drain and dry.

Counting

11 Divide the flask surface on the side with the cell sheet into eight equal areas using a ruler and thin tipped permanent marker. (This aids in keeping track of which sections have been counted.)

12 Count the total number of definite colonies and record.

Calculations

Add together the total colony count from all four flasks and divide by 4 to give the average number of colonies per flask.

$$\frac{\text{The average of the test batch}}{\text{The average of the control}} \times 100 = \text{the percentage of the control}$$

It should be noted that plating efficiency measurements are derived from counting colonies over a certain size (approx. 20 to 50 cells) and these colonies will have grown from a low inoculum of cells. This should not be confused with seeding efficiency, which is the recovery of adherent cells after seeding cells at a higher density.

Storage and handling

Serum is usually stored at −20 °C and if a large batch has been ordered, several suppliers will offer a reserve facility where serum can be held at no additional cost to the user and called off as required, if freezer space is at a premium. Serum will keep well over 12 months if stored at −20 °C and more prolonged storage may be possible at −70 °C.

Recently Sigma have advertised "cool calf serum", a specially formulated iron-supplemented calf serum which can be stored at 4 °C. Serum should not be repeatedly frozen and thawed and, if a store of serum partially thaws, e.g. if a freezer breaks down, it should be thawed completely before refreezing to minimize protein denaturation due to salt concentration effects. It is good practice to aliquot serum when thawed and keep aliquots separate for different cell lines as a safeguard against cross-contamination.

The presence of small molecular weight constituents, e.g. glucose, nucleosides, etc. may be undesirable for certain studies, in which case they may be removed by dialysis using conventional dialysis tubing and dialyse serum against Hanks's balanced salt solution (Sigma Biosciences, 1995).

Similarly, serum may be heat inactivated to inactivate adventitious agents, e.g. heterophile antibodies may interfere with cytotoxicity assays during tissue typing unless removed by heat inactivation. Also, serum may be adsorbed with charcoal to remove hormones. Although, for certain specialized work it may be necessary to have ultra pure serum, it is worth remembering that the more constituents that are removed from sera the greater the possibility of removing a vital commodity! It has been known for researchers to purify their serum to an extent where cells are actually lacking vital components for growth.

The vast majority of laboratories will purchase ready-prepared serum, which has been quality control tested as described above but, for those laboratories who find it necessary to carry out their own preparation, arrangements may be made to collect whole blood from a slaughterhouse or it maybe withdrawn from live animals. The blood should be allowed to clot at 4 °C overnight before separating the serum from the clot and centrifuging at 2000 g for 1 hour to remove the sediment. The serum may then be sterilized by filtering first through a graded series of glass-fibre prefilters and then through a final sterile 0.2 µm filter (Freshney, 1983). The serum may then be bottled in suitable-sized containers and frozen as quickly as possible. Bottles should not be filled completely to allow for the anomalous expansion of water during freezing. The usual quality control procedures should be carried out as for medium.

Suppliers of serum for batch testing

Gibco-Life Technologies: Life Technologies Ltd, 3 Fountain Drive, Inchinnan Business Park, Paisley PA4 9RF, UK   Free Phone Orders: 0800 269 210   Free fax Orders: 0800 838 380 Tel: 0141 814 6100   Fax: 0141 814 6317

Hyclone Labs Inc., 1725, South Hyclone Road, Logan, Utah 84311, USA   Tel: (US)1-800–Hyclone-492-5663   (other countries) 1-801-753-4584

PAA Biologics, Wiener Strasse 131, A-4020 Linz, Osterreich/Austria   Tel: (+43)732-49531-300; 2531, Petaluma Ave., Long Beach, CA90815, USA Tel: 310–430–9297; 9 Derwentside Business Centre, Villa Real, Consett, Co. Durham DH8 6BP, UK   Tel: 01207 582993

Sigma-Aldrich Co., Fancy Road, Poole, Dorset BH124QH, UK   Tel: (UK) (freephone) 0800 373731   (overseas) (reverse charge) 01202 733114

References

Battista, P. J., Bowen, H. J. & Gorfien, S. F. (1995). *Focus,* **17** (1), 10–13.

Biowhittaker serum supreme sales leaflet 1994. Biowhittaker UK, Biowhittaker House, 1 Ashville Way, Wokingham, Berkshire RG41 2PL, UK   Tel: 01734 795 234   Fax: 01734 795 231.

Freshney, R. I. (1983). *Culture of Animal Cells.* Chap. 10. p. 97. New York: Alan R. Liss, Inc.

Godwin, G., Danner, D. & Gorfien, S. (1995). *Focus,* **17** (1), 45-7.

Hodgson, J. (1991). Checking sources: the serum supply secret. *Biotechnology,* **9,** 1320–5.

ICN Flow Serum Terminator sales leaflet. (ICN Flow Unit 18, Thame Park Business Centre, Wenman Road, Thame, Oxfordshire OX9 3XA, UK)

Sigma Biosciences (1995). **1** (1), 2–3.

Todaro, G. J. & Green, H. (1963). Quantitative studies of the growth of mouse embryo cells in culture and their development into established lines. *J. Cell Biol.,* **17** (2), 299–313.

5

Cell culture

As explained in Chapter 1, one of the advantages of cell culture is that it may be used, in many cases, as an alternative technique to the use of live animals for studying models of physiological function *in vivo*. It is therefore essential that *in vitro* conditions mimic as closely as possible those that the cell would encounter *in vivo*. An exact replica of these conditions cannot be achieved, and consequently there will inevitably be changes in cell characteristics. Cell-to-cell interactions are reduced and cells which would not normally proliferate *in vivo* will do so under *in vitro* situations. This eventually leads to the growth of unspecialized cells rather than the expression of differentiated functions, which is why it is important to check cell characteristics regularly when cells have been cultured for any length of time. (Further discussion on this and other related factors will be covered in Chapter 8, Quality Control.)

Primary culture techniques are explained in Chapter 6. Once a primary culture has been maintained for some hours, some cell types will proliferate whilst others will survive but not multiply and others will die off. In this way, in the case of monolayer cultures, the distribution of cell types will alter until the cells are confluent, i.e. the cells are touching each other and there is no more substrate space. At confluence, the culture will have its closest resemblance to the parent tissue. After passage or subculture, the primary culture becomes a cell line (now termed a secondary culture) and may be subcultured several more times (but not necessarily established, after a few passages). The most rapidly growing cells will predominate and slow-growing cells will be diluted out. By the third passage, a culture will become more stable. Many cell lines may be grown for a limited number of passages (cell generations) after which they will either die or give rise to continuous cell lines. Some cell types, e.g. human fibroblasts will never give rise to continuous cell lines (Abercrombie & Heaysman, 1954). The ability of a cell line to grow continuously is probably related to its chromosome state, e.g. human fibroblasts remain predominantly euploid; continuous cell lines are usually aneuploid.

Most normal cells do not produce continuous cell lines but go through a 'crisis' after approximately 50 doublings after which they die out. Those which survive the 'crisis' usually produce continuous cell lines, e.g. for human diploid fibroblasts senescence will occur between 30 and 60 cell doublings or 10 to 20 weeks, depending on the doubling time.

Growing cells

It is essential that cell cultures are kept free of any form of contamination, micro-organisms present a particular danger. This being the case, anything entering the culture must be sterile (see Chapter 8 on Quality Control) and any manipulations must be carried out in such a way as to avoid the introduction of contamination.

There are a few 'golden rules' which should be observed to reduce the likelihood of contamination.

1 The work surface should be kept clean and tidy.
2 Swab the surface with 70% alcohol before starting work.
3 Keep the area as clear as possible, only introducing the items necessary for the manipulation, swabbing the outside of items with 70% alcohol where necessary.
4 Leave a reasonable space for the manipulations in the centre of the work area, and if working in a cabinet make sure this space is well inside and avoid working at the edge of the area. Working in a cluttered area increases the likelihood of a pipette touching a non-sterile object.
5 Mop up any spillages with 70% alcohol, as soon as they occur.
6 If manipulating several cell lines in succession, to avoid contamination of one cell type with another, clear up between each one and swab down the work area. Aliquot medium/serum beforehand so that each cell line has its own bottle.
7 Personal hygiene is also important and the wearing of labcoats should be adopted as regular procedure. Before beginning work, hands should be washed or gloves should be worn, preferably washed in 70% alcohol after putting on, and long hair should be tied back. Tissue culture and respiratory infections do not mix! Therefore, anyone with a cold, etc. should wear a face mask whilst carrying out manipulations unless working at a vertical laminar flow cabinet. Conversation should be avoided.
Tissue culturists must also avoid home beer-making because of the problems of inadvertently introducing yeast contamination.

8 Mouth pipetting should be avoided. This is as much of a hazard to the manipulator as it is to the cells being manipulated. Use only plugged pipettes for transferring medium.

9 If it is unavoidable, when working on the open bench, flame necks of bottles and pipettes, rotating in the flame and work preferably in an area of the laboratory where you can avoid air turbulence.

10 When removing the caps from bottles, hold the cap in the crook formed between the little finger and the heel of the hand. Alternatively, the cap may be placed downwards on a sterile surface, so ensuring that nothing will drop into it.

11 Use a new pipette for each manipulation, i.e. it is bad practice to dip in and out of bottles with a pipette that may have come into contact with cells through an aerosol.

12 The use of a suction pump with a collection trap is the quickest and safest way of removing media from flasks and dishes.

13 Never reach over an open bottle or dish, even gloves or sleeves can be a source of particulate contamination.

14 When withdrawing a pipette from its can or wrapper, take care not to touch it against any surface that may not be sterile. If in doubt, discard it.

15 Never draw up so much fluid into the pipette that it wets the plug (see Chapter 2). Anyone preparing their own pipettes should ensure the pipette mouths are plugged before sterilization.

Maintenance

Cells fall mainly into two categories, suspension cells and anchorage–dependent cells, but requirements common to both include a suitable temperature and a suitable medium with supplements at the correct pH. The temperature will depend on the species, whether the cell line is a temperature sensitive mutant, etc. For the most part, mammalian cells will grow at 37 °C. Insect and some fish cells prefer temperatures between 25 °C and 28 °C.

Most cell lines will grow well at pH 7.4 and Phenol Red is commonly used as an indicator (this will usually be included during media preparation). It is red at pH 7.4, yellowish orange at pH values lower than this, and appears progressively more purple at pH values above pH 7.4. Medium will become exhausted and acidic (yellow colour) when cells are very densely populated, indicating a need for passaging or medium changing. Some Mycoplasma contaminants, e.g. M. *arginini* will use up the arginine in the culture medium and produce an acidic pH. Conversely, cells which are not growing well will not produce much in the way of waste products, and medium will tend not to turn yellow. If there is insufficient carbon dioxide in the culture, the medium will appear purple. Other important factors concerning the environment of the cell with respect to the medium, e.g. buffering, osmolarity, etc. are covered in Chapter 3.

Carbon dioxide is essential to obtain maximum cloning efficiency for most cells (this should also be considered when seeding cells at low densities). A concentration of 5% CO_2 is usual but, if one is using only Dulbeccos' modification of Eagles' medium, this is usually equilibrated with 10% CO_2. If the CO_2 tension is altered, the bicarbonate must be adjusted so that the pH is at 7.4. Oxygen is also an essential element for growth of mammalian cells and can limit population densities particularly in large-scale cultures (see Trouble shooting).

Suspension cultures

Cells which are non-adhesive (e.g. *Ascites* tumours, many mouse leukaemias) and cells which grow continuously in suspension, e.g. human B lymphocytes transformed with Epstein–Barr virus, may be maintained by dilution or increasing the volume without subculture. (They will respond to being spun down and resuspended in new medium, but this should not be repeated too frequently.)

Suspension cells may be grown in Petri dishes (small volumes) but it is more convenient to culture them in sterile tissue culture flasks or spinner vessels. The cells should be counted, the cell suspension withdrawn, and fresh medium added to restore the cell concentration to the starting level.

Harvesting cells

1 Take the culture to a laminar flow hood.
2 Mix cell suspension and dispense any clumps with a pipette.
3 Perform a viable count (see below) using 1% Nigrosine dye (or 0.4% Trypan Blue in balanced saline.)
4 Select the desired volume and decant into sterile centrifuge pots or tubes (see Chapter 2).
5 Spin at 4 °C at 195g without any braking. For small volumes e.g. up to 100 ml, 10 minutes is usually adequate but, for 1 litre volumes, 30 minutes will be required.
6 Carefully remove the pots from the centrifuge and gently pipette off the supernatant taking care not to disrupt the pellet.

7 Resuspend pellet in the required volume of RPMI 1640 + FCS.
8 Transfer to a sterile container.

Subculturing cells

Normally carried out at 48–72 hour intervals.

1 Perform a viable count using 1% Nigrosine dye.
2 Calculate the dilution required to reduce the density of cells to 5×10^5/ml for normal suspensions or 2×10^5–3×10^5/ml for hybridomas.
If maintaining stock cultures:
   (a) remove the appropriate volume from the culture vessel and discard.
   (b) make up the volume to the desired volume, e.g. 200 ml, with a suitable medium, e.g. RPMI 1640+10% FCS. Mark the flask with date and volume and return to the incubator:

e.g. Culture vessel contains 200 ml at 9×10^5/ml
   Final volume cell concentration required=200 ml at 5×10^5/ml
   Volume of cells to be retained $= \frac{5}{9} \times 200 = 111$ ml

89 ml of cells are removed and discarded and 89 ml of fresh growth medium added.
If expanding the culture volume:
(a) Add the appropriate volume of medium to reduce the number of cells per ml to the required level, using additional cultures vessels if necessary:

e.g. Vessel contains 1 litre at 8.5×10^5/ml
   Cell concentration required=5×10^5/ml
   Dilution required=$\frac{8.5}{5} \times 1000$ ml=1.7×1000 ml

Final volume will be 1700 ml at 5×10^5/ml.

Performing a viable count

1 Mix the cell suspension and transfer 1 ml to 1 ml of 1% Nigrosine or 0.4% Trypan Blue.
2 Mix well and carefully introduce some of the mixture under the coverslip of an Improved Neubauer counting chamber using a sterile pasteur pipette. Examine under the microscope.
3 Count the number of cells that have not absorbed the dye (dead or dying cells will absorb the dye in varying amounts from the surrounding

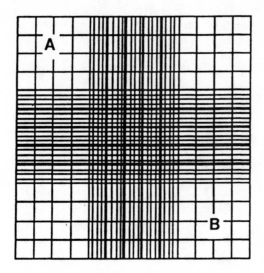

Fig 5.1. Graticule layout on Neubauer counting chamber (haemocytometer).

solution) in two complete diagonally opposite squares (see Fig. 5.1) A+B.
4 To estimate the number of cells per ml of original suspension, multiply by 10^4:*

e.g. Total number of cells in two complete corner squares=70
   Total number of cells per ml of original suspension=$70 \times 10^4=7 \times 10^5$/ml

Note

* Each corner square is 1 mm square and 0.1 mm deep and so has the volume of 1 $\times$ 1 $\times$ 0.1=0.1 mm^3. Counting two complete corner squares gives the number in 0.2 mm^3 of dye solution, and since the original suspension was diluted 1:1 with the dye, the count in two complete corner squares is the number in 0.1 mm^3 of undiluted suspension. To obtain the number of cells per ml, multiply the count obtained by 10^4 (1 ml=1 cm^3=$10^4 \times$ 0.1 mm^3)

Monolayer cultures

Once a culture has been initiated, it will require 'feeding' (periodic medium changing) and eventually subculturing. When cells are not dividing very fast, it will still be necessary to change the medium periodically as some constituents of the medium will degrade with time. Also, the cells will continue to metabolize, producing toxic substances which must be removed. There are no hard and fast rules concerning the length of time that may elapse between medium changes, this will depend on the density of the culture and the metabolic rate for each cell line. In general, for a fast growing cell line which is going to become confluent within a week of seeding, it will be necessary to feed after 2–3 days. Cell lines which are very slow growing may only require to have a change of medium weekly, though more commonly twice a week.

As an indication of whether cells require a change of medium, the pH of the culture can be observed. As the pH drops to pH 6.5, most cells will cease to grow and, as the pH drops even lower, viability will also fall and, if Phenol Red is present in the medium, these pH values will be accompanied by colour changes turning from red to orange to yellow.

At high cell concentrations normal cells, e.g. diploid fibroblasts, will stop dividing due to contact inhibition (Hayflick & Moorhead, 1961), reduced cell spreading (Stoker et al., 1968) and depletion of nutrients especially growth factors in the medium (Stoker, 1973). Normal cells will not deteriorate very much even if left at confluence for a period of two weeks. However, transformed or transfected cells and continuous cell lines will deteriorate rapidly once confluent if left without subculture or medium change. Deterioration will be apparent in the form of rounding up of cells and detachment from the surface of the dish or flask, vacuoles appearing in the cytoplasm or a granular appearance around the nuclei. Once cells have reached this stage, they are not suitable for experimental work and if the culture cannot be rescued (see Troubleshooting) the cells should be discarded and a new stock recovered. Allowing monolayer cultures to become over-confluent is not the only reason for deterioration as will be discussed later.

Subculture

It is good practice to examine cultures, with an inverted microscope, on a daily basis so that any problems may be observed and tackled at an early stage. Monolayers may be harvested by the use of a number of proteases, e.g. pronase, dispase, collagenase or most commonly trypsin. Trypsin is a proteolytic enzyme which works by digesting the proteins in the cell membrane that anchor the cell to the surface of the tissue culture vessel. Trypsin will also digest other cell membrane proteins and so it is important to keep the time it is in contact with the cells to a minimum. After the cells detach, growth medium is added. The serum proteins in this medium inhibit further action of trypsin on the cells. The enzymatic activity of the trypsin is progressively removed by heat, so that thawed bottles of trypsin should not be left for long periods of time at 37 °C. Detachment of cells with trypsin is generally more efficient if magnesium and calcium ions are removed from the medium. This can be achieved by combining the trypsin with a solution of EDTA ('versene') which is employed in a calcium and magnesium-free salt solution, i.e. 4 ml of 0.25% trypsin is added to 16 ml versene (0.2 g/l in PBSA). EDTA is a chelating agent (i.e. it will bind divalent cations) and may be used alone for the removal of weakly adherent cells. The following protocols are used in our laboratory.

Harvesting cells

1 Remove the growth medium from the culture vessel by suction and discard.
2 Add an appropriate amount of trypsin/versene or versene alone (e.g. 20 ml in a roller bottle, 10 ml per 100 mm Petri dish). Wash the fluid around gently and discard.
3 Add more trypsin/versene, wash the fluid around gently but this time discard all but a small volume (e.g. 2–4 ml per roller bottle, 0.5–2 ml per 100 mm Petri dish, 0.5–1 ml per 25 cm^2 flask, 1–2 ml per 75 cm^2 flask).
4 Examine the cell sheet at frequent intervals and tap the vessel gently to aid dispersion of the cells, incubating at 37 °C if progress is slow.
5 When the cell sheet is sufficiently dispersed, add an appropriate amount of growth medium (with serum) and carefully resuspend the cells in the medium, e.g. approximately five times the volume of trypsin left on the cells.
6 Transfer the cells and medium to a sterile bottle or flask. Add a small amount of fresh medium to rinse out the remaining cells and also transfer to the stock bottle. Perform a viable count as described above.

Reseeding monolayers

There is a wide variation in the rate at which monolayer cells divide and it is important to know the correct dilution factor for each individual cell line (split ratio). Some cells require as little as a 1:2 split, i.e. the cells harvested from one culture vessel are used to seed two vessels or half of the harvest is used to seed one vessel (of similar size). However, some cells require splitting at 1:6 or 1:10, e.g. BHK Clone 13. As a general rule, if one is not familiar with the cell line it is a good idea to subculture at several different densities for future reference.

1 A viable count is performed on monolayer cells primarily to determine the health of cells before reseeding (but it will be obvious if a culture has a low viability as dead cells will slough off from the surface and float in the medium).
2 Calculate the total number of cells needed to seed the required number of vessels, agitating gently to keep the cells in suspension.
3 Transfer open culture vessels, e.g. Petri dishes, to the CO_2 gassed incubator at the correct temperature, and closed culture vessels to a warm room or warm cabinet.

In order to assess when a cell line will require subculture, how long it will take to reach a certain density, etc., it is necessary to become familiar with the cell cycle for each cell line, as cells at different phases of the growth cycle proliferate at different rates and show differences in respiration, synthesis of a variety of products, enzyme activity etc. The growth cycle can be divided into three phases.

The lag phase

When cells are taken from a stationary culture there is a lag before growth begins, i.e. this is the time following reseeding after subculture. During this time, the cells attach to the substrate and spread out.

The log phase

This is the phase when the cells are growing and population doublings occur. The length of the log phase depends on the density at which the cells were seeded. The log phase is the optimal time for using cells for experiments as the culture is in its most reproducible form.

The plateau phase

When the culture becomes confluent, growth rate slows and eventually may stop and specialized proteins may now be synthesized faster than during other phases. Whereas in the log phase the growth is high (90–100%), this percentage drops to between 0 and 10% in the plateau phase.

The construction of a growth curve from cell counts made at intervals after subculture will determine the length of each phase of the cycle (Freshney, 1983).

It is essential for the tissue culturist to keep full and accurate records on such subjects as the origin of cells with any useful references, the tissue culture regime, i.e. dates of feeding and passaging (with a record of the passage No. and the size of the split if relevant, e.g. 1:3.)

It is also useful to have a record of the culture medium used and any additives with a note of concentrations, suppliers, etc. If changes in morphology, growth rates, etc. occur, it will be helpful for future reference to have a note of when these events happened, i.e. how long after culture initiation, etc.

Careful records are also required for the storing of cells as to how they have been stored (which method of cryopreservation has been used) and their position in liquid nitrogen vessels, so that they may easily be found again. Attention should be paid to the correct naming of cells, as there are many cell names which only differ from one another by one letter or digit.

A large number of computer programs can be designed to carry out the necessary information but a quick card reference system is also very useful.

Substrates

The choice of vessel used for tissue culture will depend on the ultimate size of the culture, the availability of vessels and a suitable place to incubate them and, in some cases, the special requirements of the cells being cultured.

From the early days of tissue culture, glass was used as the substrate because of its optical properties and because it carries the correct charge for cells to adhere and grow. Most monolayer cells require to spread out before they can begin to grow, and will not do so if they have not attached or spread (Fisher & Solursh, 1979). Glass is a cheap substrate and can be washed and sterilized without losing its growth-supporting properties. Today, the use of glass has been largely superseded by the use of plastics. Petri dishes, flasks, assay plates and roller bottles are all available commercially, and all tissue culture plastics are treated to produce a charged surface which is wettable. Plastics for tissue culture are designed for a single use, since they cannot be washed, as detergents destroy the suitability of the surface for cell adhesion. (It is worth noting that treatment of glass with strong alkali, e.g. sodium hydroxide has a similar effect but this can be reverted by treating the glass with an acid wash). Most adherent cells will proliferate on glass or tissue culture plastic, but some cells require other substrate treatments. It is noticeable that cells often grow better on a surface that has already been used and this is due to a conditioning of the surface by fibronectin (Gilchrest, Nemore & Maciag, 1980) or collagen (Kleinman et al., 1979).

Collagen improves the attachment of cells such as epithelial cells, muscle cells etc. (refer to other books in the Handbooks in Practical Animal Biology series). In our laboratory, some strains of PC12 (rat adrenal pheochromocytoma,) cells require a collagen coating to spread. The following method of treating vessels has proved useful.

Collagen preparation

Add 50 mg rat tail collagen type I (Stratech Scientific Cat no. 40236) to 500 ml of 0.1 M glacial acetic acid to obtain a 0.01% collagen solution.

Stir at room temperature for 1–3 hours until dissolved.

The collagen can be sterilized by transferring the solution to a screw-capped glass bottle and carefully adding an aliquot, approximately 10% that of the collagen, of chloroform which will form a layer beneath the collagen solution. Do not shake or stir.

After overnight incubation at 4 °C, aseptically remove the top collagen layer and store in sterile tubes at 4 °C.

Add sufficient collagen to cover the surface of the required number of tissue culture flasks and leave for 6 hours at 37 °C or overnight at 4 °C.

Remove excess fluid from the coated surface of the flask and allow to completely dry by incubating at 37 °C leaving the flask caps loose.

Prior to the addition of cells, rinse the flasks three times with phosphate-buffered saline (PBS).

(Collagen may also be prepared in the laboratory from rat tails. Rat tails are sterilized in 95% alcohol and broken to expose the tendons, which are soaked in 1:100 acetic acid and left at 4 °C for 48 hours. This is followed by centrifugation for 1 hour at 2500 g to separate the residue from the collagen in the acetic acid. The material can be stored at 4 °C for several months but should be dialysed against distilled water before use. This will result in an increase in viscosity).

Sometimes, sufficient collagen and/or fibronectin may be obtained to sustain a new culture by using the 'conditioned' or spent medium from another culture (Stampfer, Halcones & Hackett, 1980). It is essential to ensure that there is no cross-contamination of cell cultures as a result.

Feeder cells

While coating the surface of a vessel may be sufficient, it may be necessary to provide some specialized cells with a monolayer of an appropriate cell type to act as a feeder layer. The cells selected to act as a feeders must first be treated so that they will not continue to proliferate and take over the culture of desired cells. The feeder cells may be irradiated or treated with Mitomycin C (Sigma M-0503), which will halt DNA synthesis. The action of the feeders is to condition the medium and supplement it with cell products.

As a standard procedure, we use feeder cells when carrying out transformation of human B cells with Epstein–Barr virus, e.g. Use confluent Petri dishes of HFF (human foreskin fibroblasts).

Treat with Mitomycin C at a final concentration of 10 μg/ml in growth medium for 2 hours at 37 °C. (Mitomycin C is light sensitive, so the dishes may be covered with foil.) Before use, the feeder cells should be washed with PBS.

The above is not an exhaustive list of substrates and, for further information, refer either to the chapter on specialized techniques or to other books in this series directly concerned with the cell type of interest.

Daily routine

The introduction of a daily routine into the tissue culture laboratory should result in the lessening of the risk of contamination, a maintenance of workers safety and fewer problems with cell growth.

1 Check the incubators. Record the temperature and check on humidity and CO_2 levels. If there is a problem with any of these, deal with it immediately.
2 Collect anything that will be needed for tissue culture work, e.g. sterile pipettes, disposable gloves, full gas cylinders, Nigrosine or Trypan Blue for viability counts, sterile centrifuge tubes, glassware, plastic flasks, spinner bottles and Petri dishes.
3 Switch on water baths or hot plates and place medium to warm.
4 Thaw out additives, serum aliquots, etc. that will be required.
5 Wipe down bench surfaces with 70% alcohol or a commercial anti-microbial solution, e.g. Virkon (Jencons Scientific Ltd.)
6 Turn on laminar flow hood and wipe the working surface with 70% alcohol.
7 If using aspirator, check that the pump is functioning and that the aspirator is clean and contains 1% Chloros (bleach) or Virkon. If using filters or trap systems, ensure that these are clean and functional.
8 Transfer the minimum of equipment needed to the hood. It is unwise to clutter the hood with unnecessary articles as this can interfere with the air flow. Any media bottles, etc. should be wiped with 70% alcohol before introducing them into the hood.

9 If the laminar flow hood is equipped with an ultraviolet light, ensure that it is switched off. Ultraviolet light may damage cells and presents a hazard to the worker as it can result in eye damage and skin burns. The use of Bunsen burners in laminar flow hoods is not recommended as the flame disrupts the air flow. (Flame boys may be used.) However, if they are being used in a still air hood, check that the tubing between the burner and the gas supply is in good condition. Remember to keep any wash bottles of 70% alcohol well away from burner flames.

10 Check that automatic pipettors are functioning.

11 Go through all the incubators, checking all the cultures. It is worth ensuring that all operators using the same incubator will be as rigorous, if any contamination is found, it can be dealt with immediately, i.e. discard any contaminated cultures. Suspension cultures may be destroyed by the addition of Chloros (bleach) and contaminated adherent cultures may be autoclaved immediately. If cultures are observed microscopically every day, one will become familiar with the morphological appearance of different cells and will also begin to be aware of what is a usual growth rate for a particular cell line. This will lead to an ability to see when irregularities occur which, in turn, may allow one to rescue a culture at an early stage. At the end of the day, or when tissue culture work is complete, the following routine may be followed.

12 Turn off vacuum pump (and gas supply if this has been used). Remove all equipment from the hood and wipe down the interior with 70% alcohol.

13 Discard all waste material in suitable bags for autoclaving. Any living material in waste media which has been sucked off into the waste flask will be rendered inactive by the presence of the bleach (Chloros) in the flask. Check that there is sufficient bleach present and the waste flask is not over-full. (Neglecting this may result in fluid being sucked back into the pump which will be damaged.)

14 Switch off and cover microscopes with a dust cover.

15 Switch off waterbaths, etc.

16 Check that incubator doors are closed, that lids have been securely replaced on cryogenic vessels and that laminar flow hoods that have been turned off have their night doors in place.

Weekly routine

1 Clean all hoods thoroughly, removing the working surface tray so that the base is accessible. (Many of the present-day microflow hoods have

perforations which allow unobserved spillages to fall through on to the base below.)

2 Clean out all incubators thoroughly, removing shelves and paying particular attention to door seals. If a tray is being used to hold water for humidification, this should be removed, cleaned and refilled. If using a disinfectant in the tray, e.g. Roccal, this should be renewed at the specified concentration. (Many of these disinfectants damage the lining of the incubator and release toxic fumes, hence their use in culture incubators should be avoided if possible.)

3 If incubators are likely to be left unobserved, e.g. at weekends, it is wise to check the water level beforehand and top-up if necessary.

4 Ensure that all cell records are up to date.

5 Clean out waterbaths.

6 Top-up liquid nitrogen in cell freezers (this may be done twice weekly).

Troubleshooting

From time to time, particularly if a large number of different cell lines are being cultured, problems will arise with routine cell maintenance. It may be useful to consult the following checklists when considering the causes of the problem.

Contamination

This may be bacterial, fungal, mycoplasma or viral (see Chapter 8 Quality control, for detection and removal methods).

Having established that contamination is present and having dealt with it, it is worth considering how it may have occurred in the first place so that the same route of entry is avoided in future. Cell culture has the added difficulty that not all contamination problems show up immediately, particularly if infection is at low level, so time spent in making sure that the problem has been totally resolved may save much anguish in the long term.

Contamination may or may not affect speed of cell growth and, as emphasized above, regular observations of specific cultures will accustom the user to the normal growth rate for that particular cell line. It will then be evident if there is a change (either an increase or a decrease in growth rate).

Changes in growth rate are often accompanied by a change in morphology, which usually indicates either contamination with mycoplasma or another cell type or virus or a change in the cell due to ageing of the culture (senescence).

Normal cell lines usually have a finite lifespan, so that they cannot be passaged indefinitely. The process of normal cell ageing in culture is marked by a number of characteristics:-

(a) cell cycle time increases;
(b) decrease in cell saturation density;
(c) decreased adhesion to the surface on which the cells are growing;
(d) reduction in DNA synthesis;
(e) decrease in the response to growth factors;
(f) decrease in amino acid transport;
(g) increased vacuolation of cytoplasm.

Continuous cell lines (immortalized) may have an altered morphology when cultured for several months. (Swiss 3T3 cells may exhibit this phenomenon particularly if they are allowed to become quite confluent.) This change in appearance is often accompanied by anchorage independence and elevated saturation density, which would normally indicate a transformed cell line. However, the terms immortalization and transformation are not necessarily synonymous. Transformed cells also have a reduced requirement for serum and other growth factors and a shorter population doubling time. They also lose contact inhibition and are able to form colonies in soft agar. Some transformed cells will form tumours if injected into an immuno-incompetent host. Immortalized cells, in contrast, although continuably culturable are understood to be one step along the transformation process and do not usually show alterations in growth control.

Assuming that there is no contamination, and that cells have not senesced, problems of cell growth can usually be traced to errors in handling or problems with environmental conditions.

- Is there any change in procedure or equipment? e.g. incubators, warm room.
- Serum: new batch ? Check quality control tests (plating efficiency, toxicity, growth rates). Is the concentration in the medium correct ?
- Have cells been overtrypsinized or seeded at too low a density ? Were cells left too long in the plateau stage before subculture ? If the culture is in a spinner vessel, is there sufficient space between the culture and the top of the vessel? Cells may be short of oxygen.

Medium

Is the medium suitable ? Have any extra additives necessary been included, e.g. hormones, etc. Is the pH within 7.0–7.4? Is the osmolarity within the accepted range? (Manufacturers will print this on the bottle or powder packet.) Check that the osmometer is functioning correctly. If the medium was prepared from basic ingredients, check that nothing has been omitted. If the medium is water based, check that the water is satisfactory (check against bought in media, if media is prepared on site check filters and water supply systems). Have antibiotics been added at the correct concentration? Are bicarbonate levels correct ? If the medium is being prepared from basic ingredients, check the ingredients for sell-by dates, or if a new batch of any commodity has been opened, check this against previous batches. If medium is being prepared on site, check the bottling process. Is anything leaching out of the glass into the medium? Has anything changed concerning the method of washing the bottles?

Cell rescue

In general, it is wise to discard cultures which have suffered a drop in viability for whatever reason, but if the culture is precious and irreplaceable it may be possible to rescue live cells from a culture of poor viability.

For suspension cells, if there are few live cells left in a large volume of medium, it is wise to resuspend in a fresh medium with the live cells at a reasonable density, thus the volume will be greatly reduced. 'Lymphoprep' (Nycomed) or Histopaque (Sigma) is useful in the removal of dead lymphocytes from the live cells. 'Lymphoprep' is a ready-made solution of sodium metriozate and polysaccharides which will aggregate the dead cells and debris (or erythrocytes) so increasing their sedimentation rate. The sedimentation of the live lymphocytes is only slightly affected and these will form a layer further up the tube after the dead cells have sedimented. The following protocol works well in our laboratory.

Protocol

1 Centrifuge the cell suspension. Resuspend the pellet in approximately 20 ml of the original medium.
2 Place 'Lymphoprep' solution in a sterile conical centrifuge tube. Use a volume equal to half the volume of the resuspended cell suspension (e.g. 10 ml to 20 ml cell suspension).
3 Gently add the cell suspension drop by drop onto the 'Lymphoprep' by running the fluid down the tube wall.
4 Centrifuge at 1800 rpm for 20 minutes (without the brake).
5 Remove the white cloudy layer containing the viable cells from the surface of the 'Lymphoprep' and transfer to a fresh tube.

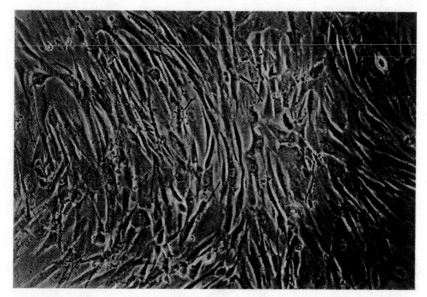

Fig. 5.2. Normal healthy human fibroblasts at early passage in log. growth. The culture is not confluent and cells are randomly orientated but beginning to show some orientation.

6 Resuspend cells in 20–30 ml RPMI + 20% FCS.
7 Centrifuge at 1000 rpm for 10 minutes.
8 Pour off the supernatant and resuspend the cells in the appropriate amount of normal growth medium to give 5×10^5 cells/ml.

Similarly, cells which adhere to the substrate should not be left to grow at too low a density. Should this situation occur following the death of the majority of cells in the vessel, the live cells should be transferred to a smaller vessel.

However, care should be taken when rescuing cultures, that one is not, in effect, carrying out a cloning of the cell line. If only one type of cell is rescued from a culture, the properties of that cell line may be lost.

Some photographs of healthy versus unhealthy cells may assist the reader to recognize some common problems (see Figs. 5.2 and 5.3).

Sometimes the worker may be presented with a sick culture from the onset. This may happen if cells have been subjected to rough treatment during transport (usually from another research laboratory since cell banks take great care in the preparation of cultures for transport). Similarly, cells may have been obtained from a laboratory where the supplier is unused to tissue culture and/or may not have sent vital information on the type of medium to use, etc. (see Chapter 3). On arrival, growing cultures should be

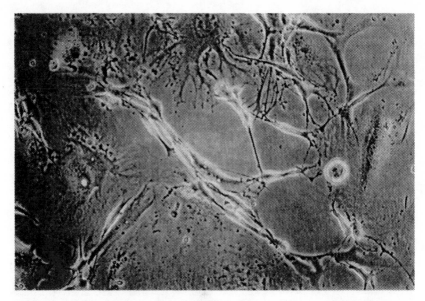

Fig. 5.3. Similar fibroblasts at late passage. Cells having varying morphology, have become granular and show total randomization. Growth rate has declined and the culture appears unhealthy.

quarantined (see Chapter 8) and left to equilibrate in an incubator at suitable temperature. It may be useful to retain most of the medium used to fill the flask during transport, then if there is any doubt as to which medium should be used, this can be re-used as an interim measure.

If a culture is not thriving, it is always worth trying a richer medium, increasing the serum concentration or increasing/renewing levels of gluta-mine. The choice of medium may not always be obvious and, indeed, selection may be made on the basis of experience rather than knowledge based on theory, but if cells are not growing well in a medium, even if it is the one recommended for that cell type, it is always worth looking for an alternative (see Chapter 3 for details), e.g. adherent cells not growing well in DMEM may pick up if RPMI 1640 is used. It should be mentioned, however, that if cells which were dividing and growing well in a particular medium begin to wane, it is likely to be due to a problem other than the choice of medium.

As a general rule, the growth of cells is improved if the culture is not too sparse so that, if a culture is found to be at too low a density, it is worth trans-ferring it to a smaller vessel, or in the case of a suspension, reducing the volume by pelleting the cells and resuspending at a greater density in less medium.

In the case of a suspension culture where there are too few live cells to be able to carry out a 'lymphopreparation', growth may be encouraged by putting the cells on to feeders in 96- or 24-well tissue culture plates.

When handling an unfamiliar cell line, it is wise to seed cells at a variety of different densities, the first time they are passaged until the worker is used to the growth rate.

Some monolayer cultures, although untransformed, tend to grow in piles rather than as an even monolayer, and there is a danger of cells at the centre becoming necrotic due to a failure to receive sufficient nutrients from the medium. It is important, in this case, to resuspend the culture at frequent intervals. Although the utmost care must be taken that cells are not sheared by being resuspended through a pipette, etc. of too narrow a bore, there are a few cell lines which may be successfully resuspended without damage, through a 21 gauge syringe needle, e.g. some strains of PC12 and Hep G2. (Piling up of cells can also occur if retraction of the cell sheet happens due to cultures being allowed to become too confluent when the cell sheet begins to peel off.)

When examining an adhesion culture, sometimes a vessel may be seen with 'bald' patches where cells have not attached to the substrate or where patches of cells have died. Very occasionally, this can be due to some problem with the substrate but, more often, insufficient volume of medium combined with the vessels not being completely level in the incubator, so that an area of the culture dries out. Another cause of empty spaces in a monolayer which is otherwise subconfluent can be attributed to uneven seeding of the culture. Even if the cells have been properly mixed, cells tend to settle unevenly, often gravitating towards the centre of the vessel. This effect can be reduced by gently tilting the freshly seeded cells from side to side and then backwards and forwards and by not stacking Petri dish trays too high in the incubator (occasionally 'bald' patches are caused by culture pathogens but these are usually more discrete patches than those caused by uneven seeding).

Cell stocks

It has to be remembered that cells are living organisms and will not always conform to a generally accepted growth rate. Consequently, tissue culture can never become a totally regimented technique. However, as far as possible, consistency and reproducibility must be aimed for when carrying out experimental work, so that cell stocks should be maintained according to a routine protocol, i.e. stocks must be seeded at identical densities and growing conditions remain unaltered. Medium should be changed and cultures passaged according to a routine schedule when they are at their healthiest, i.e. during logarithmic growth phase.

References

Abercrombie, M. & Heaysman, J. E. M. (1954). Observations on the social behaviour of cells in tissue culture II. 'Monolayering' of fibroblasts. *Exp. Cell Res.*, **6**, 293-306.

Fisher, M. & Solursh, M. (1979). The influence of the substratum on mesenchyme spreading *in-vitro*. *Exp. Cell Res.*, **123**, 1.

Freshney, R. I. (1983) *Culture of Animal Cells. A Manual of Basic Technique*. New York: Alan R. Liss Inc.

Gilchrest, B. A., Nemore, R. E. & Maciag, T. (1980). Growth of human keratinocytes on fibronectin-coated plates. *Cell Biol. Int. Rep.*, **4**, 1009-16.

Hayflick, L. & Moorhead, P. S. (1961). The serial cultivation of human diploid cell strains. *Exp. Cell Res.*, **25**, 585-621.

Kleinman, H. K., McGoodwin, E. B., Rennard, S. I. & Martin, G. R. (1979). Preparation of collagen substrates for cell attachment : Effect of collagen concentration and phosphate buffer. *Anal. Biochem.*, **94**, 308.

Stampfer, M., Halcones, R. G. & Hackett, A. J. (1980). Growth of normal human mammary cells in culture. *In Vitro*, **16**, 415-25.

Stoker, M. G. P. (1973). Role of diffusion boundary layer in contact inhibition of growth. *Nature*, **246**, 200–3.

Stoker, M., O'Neil, C., Berryman, S. & Waxman, B. (1968). Anchorage and growth regulation in normal and virus transformed cells. *Int. J. Cancer*, **3**, 683-93.

6

Preparation of primary cells

A primary culture is one started from cells, tissues or organs which are taken directly from an organism. The term primary applies until the culture is passaged for the first time, when it is called a secondary culture. As emphasized previously, at confluence the primary culture will have its closest resemblance to the parent tissue and, for this reason, early passage cells have the advantage that they will not have had time to become dedifferentiated through continuous growth *in vitro* and should still retain the properties of the parent material. In the case of some cell types, e.g. avian cells, cultures can not be maintained for more than a few passages so that the preparation of primary cultures is necessary.

Primary cultures are obtained either by disaggregating tissue mechanically, this is used most often for cultures where there is little connective tissue present (Wasley & May, 1970), or with the use of enzymes to produce a cell suspension from which some cells will adhere to a suitable surface, or by allowing cells to grow out from tissue explants.

Different tissues may require specialized techniques (see other books in this series), but there are several requirements that will apply to the preparation of most primary cultures.

1 Embryonic tissues yield more viable cells and grow more rapidly in culture than adult material.
2 The number of cells seeded per vessel should be of a much higher concentration than would be used for the subculture of an established cell line, as the proportion of cells from the tissue that will survive may be low.
3 The tissue must be chopped very finely.
4 Any enzymes used for the disaggregation of the tissue must be removed from the primary culture.

Enzymes

The common enzymes of choice are trypsin and pronase as these result in the most thorough disaggregation of cells. Collagenase and dispase are more gentle but give incomplete disaggregation. Elastase, hyaluronidase and DNase may be used in conjunction with other enzymes or, in the case of DNase, to disperse DNA from lysed cells.

In our laboratory, we regularly prepare primary cultures of rodent and chick cells and have found that the following protocols work well.

The transformation of rat embryo fibroblasts occurs as two successive steps, using polyoma virus the first step is immortalization and the second step transformation. Oncogenes can be divided into either immortalizing or transforming genes, depending on whether they either extend the lifespan of primary embryo fibroblasts or induce the cells to form transformed foci (i.e. an obvious morphological change). Whole embryo cultures are often the preferred cell type used in these assays, as primary cells provide the closest approximation of the *in vitro* state and consequently are more likely to produce genuine results as opposed to established cell lines, which due to continual culture may have altered.

Preparation of primary baby rodent kidney cells

10-day old baby mice or rats are used. The procedure will be described referring to mice only. However, the protocol is identical when rats are used.

The mice are sacrificed using CO_2 gas and culture preparation begins at once.

Protocol

1 Inside a laminar flow cabinet, the mice are totally immersed in a 10% solution of Roccal (Disinfectant, Sanofi-Winthrop Medicare) for 15 minutes.

Whilst waiting, place in the cabinet:
   (a) two separate sheets of paper towel;
   (b) three sterile beakers each containing, respectively, a pair of sterile blunt forceps, fine forceps and scissors;
   (c) an incineration bag to collect waste. Use immediately for the disposal of the bag in which the mice were delivered and gloves used in the transfer of the animals to Roccal.

2 Remove the mice from the Roccal and arrange on one of the sheets of paper towel.

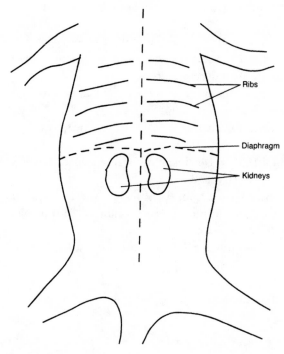

Fig. 6.1. Dorsal view showing position of kidneys, near the spine, beneath the diaphragm.

3 Take each mouse face down, in the left hand (if right handed) and position it so that the head is gripped by the little finger, the legs and tail by the thumb and the body is bent over the back of the middle three fingers

4 With the blunt forceps, take a fold of skin and peel back to expose the body wall. The kidneys are then visible near the liver, close to the spine (see Fig. 6.1). Return the forceps to the beaker.

5 Using the sharp forceps, enter the body wall above the kidneys and pull back sufficient tissue to allow the kidneys to pop out, applying gentle pressure with the hand. (It is important that the intestines are not allowed to touch the kidneys because of the risk of contamination.)

6 Transfer kidneys to a sterile glass universal bottle containing 10 ml of Dulbecco's modification of Eagle's medium (DMEM).

7 When the kidneys have been removed from all the mice, chop the material using sterile scissors until the kidney pieces are no more than 3 mm in diameter.

8 Wash three times with DMEM.

9 Add 1:1 prewarmed trypsin/versene* (approximately 10 ml per 50 kidneys)

10 Add a sterile magnetic follower and place on a hot plate stirrer at 37 °C for 5 minutes.

11 Remove from stirrer, allow tissue to settle and discard supernatant.

12 Repeat trypsinization for 10 minutes, allow tissue settle and transfer supernatant to another sterile universal bottle.

13 Add fresh trypsin/versene to the tissue and repeat until all the tissue has been digested, i.e. there are no large lumps left).

14 Add 4 ml foetal calf serum to the harvested supernatants and spin at 450 g for 10 minutes.

15 Resuspend pellets in DMEM + 10% foetal calf serum and store at 4 °C until all the cells have been harvested.

16 If Petri dishes are to be prepared from the suspension, dispense at the equivalent rate of 1.25 kidneys/60 mm dish or 2.5 kidneys/100 mm dish.

17 Dishes should be confluent in 3–4 days.

18 Dispose of mouse carcasses, paper towel, etc. Place in a sealed bag and incinerate.

* versene=0.2 g/litre
  trypsin=0.25% in Tris–saline

Preparation of whole mouse embryos

Cells are prepared from 11–15-day old embryos.
Mice are sacrificed and soaked in Roccal as described above.

Protocol

1 Remove embryos (see Fig. 6.2).
  (a) Using sterile scissors and forceps, cut the skin across the abdomen just above the hind legs. Use the forceps to peel the skin up to the head (see Fig. 6.2).
  (b) Use fresh sterile scissors and forceps to open the peritoneum, lifting the skin as you cut. Take care not to allow to foetuses to slide out of the cavity.
  (c) Using fresh instruments, carefully remove the foetuses by cutting the connective tissue and uterus and place in a 100 mm Petri dish. Wash with DMEM.
  It is important to use fresh sets of instruments for stages (a), (b) and (c) to minimize contamination of the tissue.

2 Remove the foetuses from the amniotic sacs using sterile scissors and forceps.

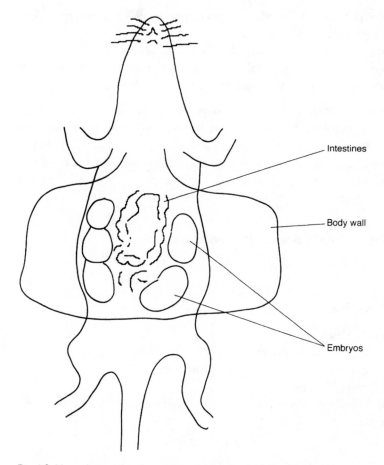

Fig. 6.2. Ventral view showing embryos contained within the uterus.

Remove heads and viscera. Transfer the embryos to a sterile container containing DMEM prewarmed to 37 °C.

3 Swill the embryos gently and discard the medium.

4 Chop the pieces into cubes (approximately 3 mm³) using fine scissors and rinse in sterile prewarmed PBS (phosphate buffered saline) until the washings are clear of blood cells.

5 Remove PBS and soak in cold trypsin/versene (0.25% trypsin in versene) overnight. Place pieces from four to five embryos in 5 ml trypsin in a universal bottle.

6 Remove the supernatant and incubate at 37 °C.

7 Add 10 ml DMEM + 10% foetal calf serum and disperse by pipette. Allow larger pieces to settle and discard.

8 Seed into 75 cm² tissue culture flasks (approximately one embryo in one to two flasks) and incubate at 37 °C.
9 Change the medium next day. (If cells are already confluent at this stage, change medium and leave for 1 hour at 37 °C before making secondary cultures or storing.) If storing, freeze 1 × 75 cm² flask to two vials.

NB In this procedure, there are many opportunities to introduce contamination. An inexperienced operator may wish to include contamination checks at different stages of the procedure. This should be done by adding samples of washings or cell suspensions to brain heart infusion broth and incubating at 37 °C. In the event of bacterial contamination, it would then be possible to show where the contamination was introduced.

Preparation of chick embryo fibroblasts

10-day old fertile eggs are used.

Growth medium

180 ml DMEM
20 ml tryptose phosphate broth } Once cell cultures
2 ml chicken serum   } become established
2 ml foetal calf serum

Work is carried out in a laminar flow cabinet using fresh sterile instruments at each step.

Protocol

1 Wash eggs with alcohol and place blunt end uppermost on an egg tray.
2 Crack top of shell with blunt forceps.
3 Remove shell to the edge of the air sac. Do not break membrane.
4 Peel back membrane, trying to avoid breaking blood vessels.
5 The embryo lies down at one side. Hook it out using curved forceps placed under the neck. Transfer to a Petri dish.
6 Remove the head. Remove limbs as near to the body as possible. Remove viscera.
7 Place the embryo into a 10 ml disposable syringe and force it through into Tris saline at 4 °C in a universal bottle, one bottle for two embryos.
8 Wash three times with Tris saline.
9 Add cold (4 °C) 0.25% trypsin in versene(10 ml/universal). Leave at 4 °C for 30 minutes, shaking every 5 minutes.

10 Allow tissue to settle. Remove supernatant to a disposable universal containing 2 ml foetal calf serum and centrifuge at 450 g for 10 minutes.
11 Add 10 ml warm 0.25% trypsin in versene to the remaining tissue and repeat the process in a 37 °C water bath until the tissue has broken down. (This usually takes approximately 30 minutes.)
12 Resuspend the cell pellets in growth medium, (DMEM + 10% FB3) mix and filter through sterile gauze into a measuring cylinder.
13 Take a small volume of suspension and dilute 1:10 for viable counting. (If there is difficulty in distinguishing embryonic cells from red blood cells the dilution can be made in 1% acetic acid, which will kill the red blood cells.)
14 Seed cells at 10^7 per 100 mm Petri dish to be confluent in 3 days. Cells may be seeded as low as 2×10^6 per 100 mm dish, but must be fed after 3 days.

Antibiotics

It is bad practice to become dependent on antibiotics because (a) when contaminations arise the organisms may already be resistant to the antibiotics used, and (b) the development of L forms of bacteria may occur. L forms of bacteria can be defined as independent growth variants (under antibiotics) of bacteria that lack a rigid cell well and have a potential reversibility to the parent strain. However, it is inevitable that antibiotics will be required at some time and the preparation of primary cultures is a technique requiring many manipulations, thus increasing the chances of contamination and may therefore be a time when the use of antibiotics is advisable.

The most generally useful antibiotic is penicillin and is usually added to media to give a final concentration of 20 to 50 units per ml. Penicillin may be used in conjunction with streptomycin (50 μg/ml) to inhibit penicillin resistant organisms.

Fungal infections may be inhibited by the use of 'fungizone' (amphotericin B) or 'nystatin' (mycostatin). It is better to keep these antifungal agents for use in an emergency and, if penicillin and streptomycin are used with any regularity, a further antibiotic should be retained for emergency use, e.g. Gentamicin since, if contamination arises in a valuable culture, it may sometimes be eliminated by the use of high concentrations of antibiotics. It should be remembered, however, that many antibiotics are toxic to cells and in some cases the effective levels are very similar to the toxic levels.

Primary cultures from explants

The growth of explants from tissues was the original technique for tissue culture and largely the only technique in use until about 1945 (Paul, 1975).

A version of an original method (Carrel and Burrows, 1912) allows a fragment of tissue to be held in place by a plasma clot and is supplied with nutrients which stimulate migration from the explant. The tissues must first be obtained and if the material is to be collected from a human biopsy there are several points to bear in mind. Usually the material will have been removed for another reason, e.g. because it is damaging to the patient or to provide material for the pathologist, etc. Therefore, when requesting samples, it is advisable to give as much information as possible as to how the sample is to be collected and to provide medium (usually the chosen medium for growth of the established culture), etc. in which to transport the material. Since samples may be collected at a time when it is not convenient to work with them, it is useful to note that most samples will survive for 24 hours at 4 °C and possibly longer, although deterioration will increase with time.

In general, the tissue should be chopped into small pieces and seeded on to a 60 mm Petri dish in a minimal volume of medium/serum 1:1 so that the surface tension will hold the pieces on the surface of the dish. Personal experience has shown that cutting tissue into small pieces with a sharp scalpel in the dish in which the culture is to be started helps the tissue to adhere to the dish as some anchorage is afforded by the slight grooves made in the plastic with the scalpel. Once this has been achieved, outgrowth of cells from the explant usually follows.

Once the pieces are seen to have cells growing out from them, the volume of medium can be increased gradually to approximately 5 ml over several days. Thereafter, the medium should be changed weekly until the cell outgrowth has substantially increased. At this stage, the explants can be removed and if desired, used again to produce more outgrowth in a new dish. The outgrowing cells should not be passaged until at least 50% of the growth surface has been covered.

Alternatively, if the explants are of very soft material, a culture may be started by pushing the material through a fine mesh sieve using the plunger from a disposable syringe. This will result in some cell debris passing through the sieve with the cell suspension, but this will be removed with subsequent changes of medium and passaging as the culture continues to proliferate.

It is advisable to freeze some vials at a very early passage so as to retain all cell types within the culture for future reference, particularly if one wishes to clone out a particular cell type in the future. (For details on freezing techniques refer to the chapter on Cryopreservation.) However, some primary cell types will not survive cryopreservation.

Many different primary cell types may be cultured by choosing the correct techniques (Refer to the appropriate book in this series) but it should be noted that, once a particular cell type has been segregated, a full characterization should be carried out for that particular cell type and it may be necessary to clone the culture at an early stage if a pure cell type is required, e.g. if setting up a primary bone culture, it would be necessary to clone to achieve a pure culture of osteoblasts.

References

Carrel, A. & Burrows, M. T. (1912). Cultivation of adult tissue and organs outside the body. *J. Am. Med. Assoc.*, **55**, 1379–81.

Paul, J. (1975). *Cell and Tissue Culture*, Chap. 11, p. 175. 5th edn, Churchill Livingstone.

Wasley, G. D. & May, J. W. (1970). *Animal Cell Culture Methods*, Chap. 10, p. 105. Oxford and Edinburgh: Blackwell Scientific Publications.

7

Cryopreservation

It is quite often the case that particular lines of cells used by a laboratory are not required in culture continuously, but it can be important to have ready access to fresh stocks when required. Cell lines can be purchased direct from national collections but this takes time and could be expensive depending on the number purchased annually. Supplies from fellow researchers in other laboratories may not be readily available and may be of unknown quality. Most laboratories involved in cell culture establish cell banks of their own in liquid nitrogen refrigerators providing instant access when required.

Establishing a cell bank

A cell bank is very simple to set up and requires the following:

1 a liquid nitrogen refrigerator or −150 °C ultra-low freezer. Liquid nitrogen refrigerators come in varying sizes from small units holding cryotubes to very large units holding many thousands of tubes (see Chapter 2 for details of suppliers);
2 storage racks and boxes in which to place the tubes;
3 a regular source of liquid nitrogen;
4 insulated gloves and a visor;
5 a supply of cryotubes. These are thick-walled screw-capped polypropylene tubes, specifically designed for liquid nitrogen storage and available in various capacities from 1.0 ml upwards. *Never* use any other type of plastic tube for this purpose;
6 a comprehensive and accurate record system for logging every cell line stored in it. This can be either computer based (with or without hard copies) or handwritten in folders.

Every time cells are put into or removed from storage, the record should be updated as soon as possible. Records should consist of the following information:

(a) location of cryotubes (cane or rack and box number);
(b) cell type name;
(c) date of storage;
(d) number of cells stored;
(e) type of freezing mixture;
(f) volume of medium to which the cells should be recovered;
(g) date cells recovered.

Freezing cells: principles

Storage in liquid nitrogen at -196 °C allows cells to be maintained in a viable state but in suspended animation, almost indefinitely. This minimizes those inherent problems of cell culture: cross-contamination between cell lines; microbial contamination; and genetic drift. Liquid nitrogen refrigerators need to be monitored routinely and fresh liquid nitrogen added as required. Different makes of refrigerator have different evaporation rates and it is important to be aware of the average loss rate and the frequency at which they will need topping-up. The loss rate is also affected by the frequency the refrigerator is opened when cells are placed into or removed from storage. A recently introduced alternative to liquid nitrogen storage is the -150 °C ultra-low freezer from Revco and Sanyo who claim that it ensures more effective long-term sample preservation than storage in liquid nitrogen. The physical conditions prevailing when cells are frozen are crucial to the success of the operation. Freezing is essentially a dehydration process. As ice crystals separate out the concentration of solutes in the cell increases and substances normally harmless may rise in concentration to toxic levels, particularly affecting the lipoproteins of cell membranes. Also, as buffering salts precipitate, changes in pH occur. The cells need to be protected from these destructive effects of ice crystal formation which, in unprotected cells, occur between 0 °C. and approximately -20 °C (the eutectic point). The eutectic point is the temperature at which ice crystal growth ceases. To achieve the highest possible degree of protection for the cells, two precautions are normally taken to avoid the potential damage outlined above. These are the use of protective agents (cryopreservatives) which suppress the eutectic point and controlled rate freezing and thawing. Controlled rate freezing and thawing are important as they fix the time of exposure of cells to high salt concentrations and the site and rate of ice crystal formation. The cells are suspended in growth medium or serum containing a cryopreservative, the most commonly used of which are dimethylsulphoxide (DMSO) and glycerol. A completely defined freezing mixture that does not contain any serum specifically designed for serum-free cell cultures is also available (ICN

Table 7.1 *Common types of freezing mixtures*

Ingredient	Percentage of total				
Medium	70	85	80	–	–
Serum	20	10	10	90	95
DMSO or glycerol	<u>10</u>	<u>5</u>	<u>10</u>	<u>10</u>	<u>5</u>
	100	100	100	100	100

Note:
A completely defined cryopreservation medium that does not contain either serum or DMSO suitable for use with serum-free cultures is available from ICN Biomedicals ('CellVation')

Biomedicals 'CellVation') (see Table 7.1 for a summary of the most common freezing mixtures used in different laboratories). Other protective agents have also been used in the past, e.g. glucose, sucrose, polyvinylpyrrolidone (PVP), etc., but DMSO remains the protective agent of choice because its eutectic temperature is particularly low (-130 °C) and its capacity for lowering the freezing point of aqueous solutions is remarkable. The small molecular size of DMSO enables it to pass in and out of living cells with less osmotic disturbance than is exerted by glycerol. DMSO and glycerol exert their protective action by reducing the concentration of salt in equilibrium with ice at any temperature, decreasing the likelihood of denaturation of proteins and other complex organic substances in the external and internal membranes and organelles of living cells. They are non-electrolytes, which bind water through lone-pair electrons on oxygen atoms. The lone-pair electrons attract the hydrogen atom in the water.

Freezing cells: practice

The candidate cells should be in the middle log phase of growth with a viability in excess of 95%. Adherent cells which are just sub-confluent should be removed from the substrate on which they are growing in the manner described in Chapter 5 and then sedimented by centrifugation to form a pellet. Similarly, suspension cells should be decanted into a centrifuge tube or Universal bottle and centrifuged to form a pellet. The medium is then decanted from the tube or bottle and the cells are resuspended in the freezing mixture of choice. It is important that the cells are stored at a concentration of 1×10^7 cells per ml or greater. After gentle mixing, the cells are usually dispensed in 1 ml or 2 ml aliquots into cryotubes and labelled clearly with the name of the cell line, the number of cells per vial, the volume to which they should be recovered, the date of freezing and, if there is room on the label, any other information considered relevant. Between room temperature and approximately -20 °C, the cells must be cooled slowly at a maximum cooling rate of 1–2 °C per minute. Between -20 °C and -130 °C the cells can be frozen more rapidly at rate of 4–5 °C per minute without any deleterious effect. From this point they can be transferred directly to the liquid nitrogen refrigerator. In practice, except where using a programmable, controlled-rate freezing apparatus, cells are placed at -70 °C/-80 °C to cool at a rate of 1 °C/2 °C per minute overnight and then transferred to the liquid nitrogen refrigerator the next day. In theory, some damage will occur to the cells but, in practice, this is sufficiently minimal not to be a problem. *Always remember to wear long-cuffed, insulated gloves and a face visor when handling liquid nitrogen.* There are a number of methods for achieving the required cooling rate. These range from the very basic expanded polystyrene box or drinking cup lined with tissue paper method to sophisticated commercial programmable cell freezing apparatus. There are now several relatively inexpensive commercially available developments of the polystyrene box method that give acceptably reproducible results. The use of this type of freezer is strongly recommended. A more detailed description of these items of equipment can be found in Chapter 2, along with a list of suppliers of cryotubes, liquid nitrogen, liquid nitrogen refrigerators and cell freezers. It is recommended that a day or two after a cell line has been frozen, a cryotube is removed from storage and a test thaw carried out to determine the success of the freeze. This should be done before discarding the stock culture in case any problems are encountered, requiring the cells to be refrozen.

Thawing frozen cells

Thawing (resuscitation) of cells stored in liquid nitrogen should be accomplished as rapidly as possible using a 37 °C water bath to avoid microregions of high salt concentration that may set up adverse osmotic conditions that could lead to rupture of the cell. Refer to Chapter 2 for important comments on the use of waterbaths. *Always wear long-cuffed, insulated gloves and a face visor when removing cryotubes from the liquid nitrogen refrigerator.* On removal from the refrigerator, the label should be checked to confirm the cryotube contains the required cell line before thawing. When the cells have thawed, mix them gently by inversion and wipe around the cap with 70% alcohol. Post-resuscitation techniques vary widely between laboratories. For example, with adherent cells the cells can either be transferred directly to a culture flask (pregassed if necessary) containing an appropriate amount of growth medium and left undisturbed for 24–48 hours before changing the medium except for a check that the cells have adhered to the vessel surface; or the medium is changed after 2–4 hours to remove the cryopreservative, dead cells and cell debris. Alternatively, the cells can be removed from the cryopreservative by centrifugation and resuspended in fresh medium for seeding a flask. With suspension cells, they can again either be left for 24–48 hours to recover before handling further; or the cells are added to 20 ml of growth medium in a sterile Universal container and spun out to remove the cryopreservative, etc. They are then resuspended in fresh growth medium and transferred to a culture vessel. A viable count can be performed on suspension cells a few hours after resuscitation to determine the health and viability of the cells. This has the advantage that, if the survival of the cells is in doubt, a replacement culture can be established the same day if necessary.

8

Quality control

Guidelines have been given in Chapters 1, 3 and 5 concerning the handling of new cell lines in such a way that the researcher can be assured that the cells are free of contamination, but quality control is not a procedure kept solely for use when new material is received or when primary cultures are set up in the laboratory. The continuous monitoring of cultures and ingredients is essential and requires constant vigilance. Some contaminants are visible to the naked eye but others, e.g. contamination with some strains of mycoplasma are not, and even the effects of such infection may not be immediately obvious, hence the need for constant testing. The importance of carrying out experiments with contaminant-free cells cannot be over-stressed (Mowles & Doyle, 1990).

Once contamination is present in a culture, it can easily be spread, so it is necessary that all staff are aware of the potential for problems to occur. However, in some cases the presence of a contaminant may not be such a calamity as it was before the days of mycoplasma removal agents and a wide range of efficient antibiotics.

Many laboratories indulge in the practice of using antibiotics in cell culture as a routine procedure. This leads to the suppression of bacterial contamination which can encourage the spread of antibiotic-resistant strains. Antibiotics can also reduce levels of mycoplasma, making them harder to detect. Microbial quality control is concerned with the testing of cell lines and media for a variety of micro-organisms including bacteria, yeasts, fungi, viruses and mycoplasma.

Bacteria, yeasts and fungi

Contamination by bacteria, yeasts or fungi is often obvious in cell culture by the appearance of turbidity in the medium or by a change in pH (which will be seen if the medium contains an indicator). Aside from the operator and cross-contamination from an already-infected cell line, the reagents added during the preparation of the culture media are the most likely route of entry of contamination so that it is good practice to set up quality control checks on culture medium prior to use. For example, test any new batch of media for growth of contaminants by incubating an aliquot at 37 °C for a minimum of 48 h prior to adding to cell cultures. Methods used for the detection of bacteria and fungi include the culture of the microorganisms in broth and direct observation using Gram's stain. This will detect bacteria, both rods and micrococci.

Gram's strain

Reagents

Crystal violet 'Gurr' 0.5% solution	Product No. 35083
Lugol's iodine solution 'Gurr'	Product No. 35088
Neutral red (aqueous 1%) 'Gurr'	Product No. 35154
Carbol fuchsin	Product No. 35159
Safranin O	Product No. 35153
DePeX mounting medium 'Gurr'	Product No. 36125

All available from BDH, PO Box 15, Freshwater Road, Dagenham, Essex RM8 1RF Tel: 0181 597 7591

N.B. Slides used must be grease free, preferably soaked in alcohol and dried with a clean lint-free tissue.

Preparation of sample

1 Transfer no more than 50 μl of sample to the central area of a slide using a pasteur pipette or bacteriological loop. Distribute the sample to cover an area approximately 2.5–3.0 cm by 1.0–1.5 cm.
2 Allow the sample to dry in air or the process can be speeded up by the application of *gentle* heat.
3 When the sample is dry, fix it to the slide by passing it through the flame of a Bunsen two or three times, using forceps to hold the slide.

Staining procedure

Slides should be placed onto a staining rack placed over a sink.

4 Cover the slide with 0.5% Crystal Violet stain. Allow to act for 1 minute.

5 Using forceps, tip the slide from the end and wash off the violet stain with Lugol's iodine solution. Return the slide to the horizontal and cover it with more Lugol's iodine and leave to act for 2 minutes.

6 Using forceps, again tip the slide from the end and decolorize the sample by washing under the running tap until decolorization is judged complete.

7 Replace the slide on the staining rack and cover with either 1% Neutral Red, Carbol Fuchsin or Safranin O. Allow to act for 4 minutes.

8 Wash off the red dye with running tap water, carefully blot dry and allow to air dry.

9 Either examine immediately under the microscope or mount a coverslip with DePeX and leave to dry before examining.

Detection by cultivation

1 Add 1 ml of test material into each of a pair of universals containing 15 ml of the appropriate broth: Sabouraud broth, which will support fungal growth, or Brain heart infusion and Tryptose phosphate broth which will support the growth of aerobic bacteria and yeast. Thioglycollate broth may be used to detect aerobic and anaerobic forms of bacteria (see Table 8.1).

2 Incubate one of each set at 37 °C and the other at 25 °C. The broths will usually become turbid within 1 or 2 days, but if infection is at a low level incubation may take as long as two weeks.

Blood agar plates may also be used to detect aerobic and anaerobic bacteria. In this case, a sample of the test material should be streaked onto plates which are then inverted and incubated at 37 °C. (Some photographs of cultures contaminated with bacteria and fungi are included), see Figs. 8.1 and 8.2.

If a cell culture is contaminated with bacteria or fungi, the best method of elimination is to discard the culture and start again with fresh stocks from a refutable source. As well as eliminating the contaminant, it is important to discover the source if possible, to prevent contamination recurring. In the case of irreplaceable stocks, it is worth trying to eliminate contamination using antibiotics (Doyle & Bolton, 1994). If carrying out such treatment in a flow cabinet, it may be helpful to work with the fan off to reduce the spread of spores from fungal contamination.

1 Culture cells in the presence of the chosen antibiotic for 10–14 days or at least three passages. If the cells are heavily contaminated with bacteria or

Table 8.1. *Broths used for the detection of fungi, yeast and bacteria*

Broth	Concentration	Organisms detected
Brain heart infusion	37 g/litre of H_2O	Streptococci, pneumococci and meningococci, etc.
Thioglycollate medium	29.5 g/litre of H_2O	Anaerobic, microaerophilic and aerobic micro-organisms
Sabouraud broth	30 g/litre of H_2O	*Aspergillus niger, Candida albicans, Escherichia coli, Lactobacillus casei, Saccharomyces cerevisiae*
Tryptose phosphate broth	29.5 g/litre of H_2O	*Neisseria meningitidis, Staphylococcus epidermis, Streptococcus pneumoniae, Streptococcus pyogenes*, etc

Notes:
All broths must be sterilized before use.
All powders are available from:
Difco Labs, Detroit, Michigan 48232, USA or PO Box 14B, Central Avenue, West Moseley, Surrey KT8 2SE, UK.

fungi, change the medium before treatment. At every passage, seed cells at the lowest density at which the cells will grow.

2 If the contaminant seems to have been eradicated, re-test the cells after culture in antibiotic-free medium after 1 week.

It is possible to retreat a bacterially contaminated culture with alternative antibiotics if success is not achieved with the first antibiotic selected, but it may be useful to carry out a Gram's stain on a sample of contaminated cells and /or a sensitivity test using antibiotic discs★ on agar plates on which the bacteria contaminant has been cultured. These tests will assist in the selection of the appropriate antibiotic.

★Available from: Unipath, Wade Road, Basingstoke, Hampshire RG248PW, UK   Tel: 01256 841144

Useful antibiotics for use against bacteria and fungi

Amphotericin B (Fungizone):	2.5 mg/litre active against yeast and other fungi.
Gentamicin:	50 mg/litre active against Gram-positive and Gram-negative bacteria.

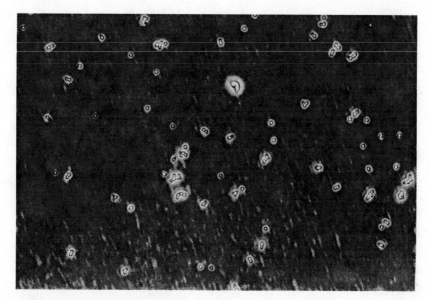

Fig. 8.1. Yeast (magnification ×270).

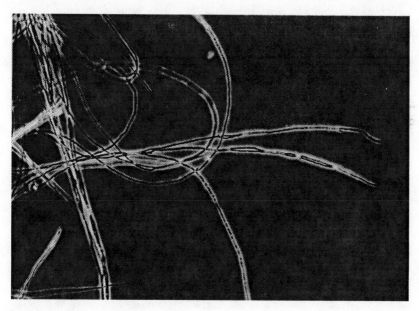

Fig. 8.2. Fungus (magnification ×96).

Nystatin: 50 mg/litre active against yeasts and other fungi.

Penicillin-G: 1×10^5U/litre active against Gram-positive bacteria.

Streptomycin sulphate: 100 mg/litre active against Gram-positive and Gram-negative bacteria.

Viruses

When considering viral contamination, as with bacterial and fungal infections, one must be aware not only of the problems produced in respect of experimental work but also the safety of laboratory staff. The source of viral contamination can be from the original tissue used to prepare the cell line, from other infected cultures, from growth medium or (in rare instances) from laboratory personnel.

Methods of detection

There is no single method for the detection of all viruses and exhaustive screening for viral contaminants is usually uneconomic. Generally, it is not possible to see the effects of viral contamination under the microscope but there are a few exceptions which may prove useful, e.g. avian fibroblasts infected with a virus such as Rous sarcoma virus will show a change in morphology. (The pattern of narrow strap-like, evenly swirling, cells will change to a disorganized culture with many rounded cells piling up on one another.)

Co-cultivation

An extract of the cell line to be tested may be incubated with semi-confluent monolayers of a range of cell lines susceptible to a wide variety of viruses. The co-cultivations are passaged for 3 weeks and the host cell lines checked for cytopathic effects. Useful host cell lines to use include BHK21 (hamster), WI38 (human), Vero (monkey) and MDCK (canine). Fresh T-cells are sensitive to HTLV-1. Some leukaemia viruses induce syncitial plaques and the best-known test for these is the XC plaque assay for murine leukaemia virus. When MLV-infected mouse cells are co-cultivated with XC cells, the XC cells form syncitia (Rowe, Pugh & Hartley, 1970). This assay will detect ecotropic MuLV viruses (ecotropic viruses infect other rodent cells), xenotropic MuLV(infects only cells other than rodent), or amphotropic viruses (infect both). Xenotropic virus may be detected by the formation of foci in

S+L− mink cells. This cell line was initiated from a single cell of a single focus produced in mink cells infected with a terminal dilution of MSV beyond the endpoint of the associated helper RD-114 virus. The S+L− mink cells contain the MSV genome rescuable by compatible helpers (Peebles, 1975).

Rous sarcoma virus will produce discrete colonies or foci of transformed cells in a culture of fibroblasts, and this property has been used to design a quantitative assay (Temin & Rubin, 1958)

Electron microscopy

Electron microscopy may be used to detect and identify viruses by their characteristic morphology.

Reverse transcriptase for retrovirus detection

When virus is released into the culture medium of infected cells, it can be detected by assaying for the presence of the viral enzyme, reverse transcriptase (Baltimore, 1970; Temin & Mizutani, 1970).

PCR

Polymerase chain reaction techniques have been devised to provide rapid results (Hertig et al., 1991) but uncharacterized virus strains may be missed by PCR owing to inadequate hybridization of primers. There are no reliable methods for the eradication of viruses, other than cloning and this raises other problems (see Troubleshooting section of Chapter 5).

Mycoplasma

Mycoplasmas are frequently found, serious contaminants of cell cultures. They may originate from a variety of sources such as the tissue used to establish primary cultures, untested serum *added to the medium or from the personnel who handle the cultures, but the most common route of infection is between one culture and another. The effects of mycoplasma contamination are various. These organisms can cause alterations in cell structure, function, metabolism, karyotope and growth characteristics and may cause cell lysis.

Since mycoplasma infections cannot be seen when observing cell cultures under the microscope (although some of the effects may produce warning signals such as acid medium, toxicity, etc.), it is essential to check regularly for the presence of the organisms. The growth of mycoplasma in cell cultures can be detected either by a direct microbiological procedure or by indirect methods which include staining, biochemical methods, hybridization techniques or PCR. Some regulatory bodies insist that mycoplasma detection is carried out by a particular method so it is worth checking if one is working under FDA guidelines, etc.

* All reputable companies test serum for mycoplasma and will issue documentation relating to the degree of testing carried out. PAA laboratories produce a serum 'mycoplex' which has been treated with ultra-violet light in order to destroy any mycoplasma that otherwise could have passed through a 0.1 μ filter.

Perhaps the most well-known method for mycoplasma testing is the process of DNA staining with dyes like DAPI(4',6-diamidine-2'-phenylindole dihydrochloride) or the more widely used Hoechst 33258 DNA staining method. (The fluorochrome dye binds specifically to DNA causing fluorescence when viewed under ultra-violet light.)

Method for mycoplasma detection–fluorescent stain

Reagents

Mitomycin C   *Sigma M0503*
Dissolve 2 mg vial in 2 ml distilled water.
Carnoy's Fixative is made up of three parts absolute ethyl alcohol + one part glacial acetic acid.

Bisbenzamide (Hoechst 33258) stain   *Sigma B2883*
Make stock solution of 100 μg/ml in PBS. This can be stored at 4 °C for up to 6 months.
Working solution is made by diluting the stock solution 100 × with methanol.

Hydromount National Diagnostics HS106

We routinely use one of two methods:
(a) the non-fixing method – This is used most frequently;
(b) the fixing method – This is used if, for any reason, staining cannot be carried out when the cells are at the desired confluence. The cells are then fixed and stained at a later date.

Non-fixing method

1 If testing a *monolayer* cell line, plate the cell line on to 35 mm tissue culture dishes to be 50–70% confluent in 2–3 days. Miss out Steps 2,3 and 4 below, continue method from Step 5.
2 If testing a *suspension* cell line, a cell supernatant or a swab sample, an indicator cell line is used. Vero cells (kidney from African Green Monkey) are good as they have a large cytoplasmic area, thus making it easier to see any mycoplasma after infection from a sample.
3 Vero cells are prepared as follows.

Grow to confluence in DMEM, 10% calf serum (CS) in 175 cm^2 flasks.
Treat with mitomycin C to stop cell division. Add mitomycin C to the flasks
    at a final concentration of 5 μg/ml medium.
Incubate at 37 °C for 2 h.
Harvest cells in normal way with trypsin/versene.
Store 1/2 flask per ampoule in a freezing mix of 90% FCS+10% DMSO.
When required, 1 ampoule is recovered into 40 × 35 mm dishes in *DMEM
minus antibiotics*+10% FCS, at 1.5 ml/dish. Incubate at 37 °C.
Once the cells have adhered these dishes can be used at any time. They
    generally last 2–3 weeks.

4 Inoculate each test sample into a dish of prepared Vero cells. Incubate at 37 °C for 3 days. Set up positive and negative control dishes using a known positive cell line and Vero alone, respectively. (Any mycoplasma work is best carried out in an area away from the main tissue culture laboratory as mycoplasma is very easily passed from one cell culture to another. If a separate area is not available, it may be worth considering having samples tested by a cell bank rather than run the risks associated with handling mycoplasma contaminated cells in the laboratory.)
5 After 3 days check all dishes under light microscope and discard any which have become contaminated.
6 Pour off medium from dishes. Wash with PBS and drain on paper.
7 Cover cells with bisbenzamide – methanol staining mix. Incubate at 37 °C for 15 minutes.
8 Pour off staining solution and drain on paper.
   NB The staining mixture is highly toxic, and disposal is by incineration.
9 Put a drop of Hydromount on slide surface and cover with a coverslip.
10 Examine under fluorescence microscope using oil immersion objective, 340/380 nm excitation filter and 430–450 nm barrier filter.

Fixing method

1 Carry out steps 1–5 in non-fixing method above.
2 Leaving medium in the 35 mm dishes, slowly add an equal volume of Carnoy's fixative and allow to stand for 5 minutes.
3 Decant, add 2 ml neat fixative and leave a further 3 minutes.
4 Pour off fixative and invert dishes on paper to drain.
   NB Dishes may be stored at this stage for several weeks.
5 Add 2 ml bisbenzamide working solution and incubate at 37 °C for 15 minutes.
6 Pour off staining solution and rinse with distilled water (collect stain and washings for incineration). Invert on paper to drain.
7 Complete method with Steps 9 and 10 above.

Another staining method utilizes Aceto-Orcein stain, and a technique used for looking at the chromosomes of a cell (Fogh & Fogh, 1964).

Reagents

1 Sodium citrate solution, 0.6%
   Sodium citrate ($Na_3C_6H_5O_7 . 2H_2O$) 6.84 g
   distilled water 1 litre
2 Carnoy's fixative
   Glacial acetic acid   1 part
   Absolute ethyl alcohol   3 parts
3 Orcein stain*

* Orcein acetic solution (La Cour) 'Gurr' Prod. No. 35090 from BDH.

Method

(Cells for test should be seeded on cover slips or Petri dishes at sub-confluent density)

1 Decant medium from cover slips after removing a small amount for agar stabs. Place cover slip in plastic well, or, if supplied in Petri dishes, treat *in situ*.
2. Add 3 ml 0.6% sodium citrate solution/50 mm dish.
3 Add 1 ml distilled water dropwise to bring the sodium citrate concentration to 0.45%.
4 Leave for 10 minutes.

5 Slowly add an equal volume (4 ml) Carnoy's fixative.

6 Pour off the fluid and replace with 2 ml Carnoy's fixative.

7 Leave for 10 minutes.

8 Suck off.

9 Allow cover slips to dry in the air for 5 minutes, or until absolutely dry.

10 Add orcein stain. Leave for 7 minutes.

11 Wash three times with 95% alcohol. Dry.

or, if using DePeX (BDH) mounting medium, wash three times with absolute alcohol, clear in xylene and mount.

12 Mount coverslips inverted in Euparal* and examine under the microscope – (× 40 objective or × 1000, oil immersion).

* GBI Laboratories Ltd, Shepley Industrial Estate, Audenshaw, Manchester M34 5DW, UK. Tel: 0161 336 5418

Mycoplasma stain black

Although the staining methods produce a rapid result, there are disadvantages. It may be difficult to differentiate between contaminated and uncontaminated cells which have a small amount of cytoplasm compared with the size of the nucleus. This is why it is a good idea to use a monolayer culture on to which the test sample is inoculated as an indicator. This method can also be used to test supernatants, reagents, etc., which do not contain cells. Contaminating bacteria or fungi will also stain, although they will be much larger (Mycoplasmata range from 0.3-0.5 μm in diameter.)

Microscopic methods require a trained eye for the interpretation of results and, in the case of the Hoechst stain, a fluorescence microscope. These methods are less sensitive than the culture method. It is generally considered that, to obtain a definite positive result using the Hoechst stain, a culture needs to have an infection of approximately 10^4 mycoplasma per ml.

Growth of mycoplasma colonies on specialized agar produces less problems with the exception of the growth of some *M. hyorhinis* strains. It was the practice in our laboratory to use a method refined by W. House (House & Wadell, 1967) until the problem of growing *M.hyorhinis* became evident. For an extended period, the agar growth method was backed up with the Boehringer Elisa assay (see below) until the discovery of an agar which would support the growth of all strains of mycoplasma. The method which follows is extremely efficient and easy to use, the only disadvantage being a slightly longer time period before results are seen. From time to time, we have compared this protocol with alternative methods/new kits on the market for mycoplasma testing and have always found this protocol to be as efficient in revealing mycoplasma positive samples. This includes a recent blind comparison with the PCR method (M. A. Harrison & D. Bicknell 1995, unpublished results.)

Method for detection of mycoplasma by growth of colonies on agar

1 Prepare an appropriate number of confluent 35 mm cell culture dishes of BHKC13 feeder cells. (The feeder cells appear to have a conditioning effect, resulting in colonies appearing 24–36 hours earlier than if no feeder cells are used.)
2 Check for degree of confluency and health of cells before use.
3 Drain the medium from the dishes and overlay with agar previously melted at 50 °C★ 2–3 ml/dish. Leave to set.
4 When the agar is thoroughly set store the dishes inverted in an airtight plastic container at 37 °C. (Dishes may be used for up to 1 week after preparation.)
5 Inoculate a dish by taking up a sample into a Pasteur pipette and stabbing the sample vertically through the agar several times. (For monolayers it will be necessary to scrape off some cells first.)
6 Allow the agar plates to dry before taping the lid on to the dish and inverting it.
7 Incubate at 37 °C in an atmosphere of 5% CO_2 in N_2 in an environmental chamber.
8 Examine microscopically every other day for up to 14 days. (Most mycoplasma positive samples (see Figs. 8.3 and 8.4) will be seen to have colonies long before this.)

★ Mycoplasma agar and supplements are available from:
Mycoplasma experience, 1 Norbury Road, Reigate, Surrey RH2 9BY, UK   Tel: 01737 22662. Fax: 01737 224751

Elisa detection of mycoplasma

The mycoplasma detection kit (Boehringer Mannheim 1296 744) will detect the most common mycoplasma/acholeplasma species contaminating mammalian cell cultures, i.e. *M.arginini, M.hyorhinis, A.laidlawii, M.orale.* The basics of the test are as follows:

The determination of each species is carried out separately. A coating antibody is fixed to a 96-well microtitre plate, and non-specific binding sites are blocked. The sample is then incubated so that the antigen (mycoplasma) binds to the corresponding coating antibody. The fixed antigen is then marked by the biotin-labelled antibody and the conjugate visualized by way

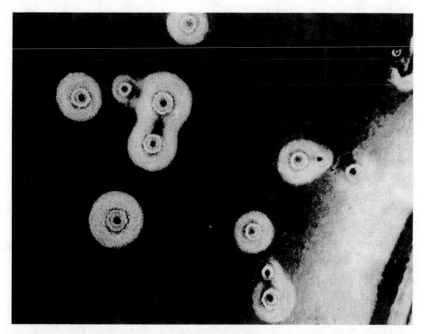

Fig. 8.3. Mycoplasma colonies growing on agar (not visible in cell culture).

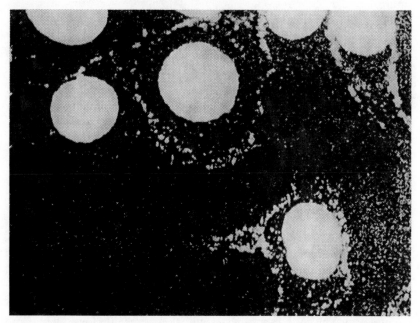

Fig. 8.4. Mycoplasmal contamination as seen when stained with Hoechst 33258.

of the binding of streptavidin alkaline phosphatase and its enzymatic reaction with 4-nitrophenylphosphate substrate. The process takes a day, with an overnight incubation, and is simple and efficient, the only disadvantages being cost and the fact that only the most common mycoplasma will be detected. It does have the advantage of identifying the species of the contaminant.

PCR

PCR methods have been developed for the rapid detection of mycoplasma in cell cultures, involving the use of universal primers for amplification of the regions between the 16S and 23S rRNA genes of the main mycoplasma species contaminating cell culture. This test can detect at least 10^3 c.f.u/ml. There are kits available from Stratagene and Boehringer Mannheim. The latter (Mycoplasma PCR elisa) is a photometric enzyme immunoassay for the detection of PCR-amplified DNA of mycoplasma/acholeplasma in cell culture.

PCR methods are ultrasensitive but require a DNA thermal cycler and are best performed by a careful worker who has space to prepare samples for test in an area away from the actual testing, as there is a high risk of cross-contamination and false positives.

The above list of methods for mycoplasma testing with their advantages and disadvantages may prove a confusing choice for the uninitiated worker. In order to select the method appropriate for your laboratory, the following considerations should be made in order of importance:-

Adherence to any set regulations by any governing body, availability of equipment, cost, time, type and number of samples to be tested. The fact that there is such a number of alternatives should impress upon the reader the importance of having mycoplasma free cultures. Mycoplasma is probably the most serious contaminant of tissue culture in that it has the most far-reaching effects.

Having established that there is a mycoplasma contamination, the following action may assist when contemplating a clean-up operation.

1 Discard all contaminated cultures (destroy by autoclaving).
2 Quarantine all suspect cultures and those precious cultures which are to be subjected to mycoplasma eradication (see below).
3 Discard or autoclave anything possible (i.e. small pieces of equipment, etc.).
4 Swab down all surfaces in the area with Tegodyne* (followed by IMS to

remove the Tegodyne. (This includes rims of pipette drums, incubator interiors, hoods, pipette aids, drawer handles, etc.)

5 Send all laboratory coats to the laundry.

6 When recovering new cells, even if coming from a reputable source, this should be carried out in an area separate from the main working area, and cell cultures should only be introduced into the main laboratory when proved to be free of mycoplasma.

7 Be aware that stocks may have been stored before it was known that there was a contamination problem. If in doubt, recover these in a quarantine area and remember to make a note in the records.

* Tegodyne, also known as Tegodor or Wescodyne, is available from:
Synchem (UK) Ltd, PO Box 170, Staines, Middlesex TW18 4UZ, UK   Tel: 01784 449966.
Fax: 01784 466207

Other methods

Nucleic acid hybridization assays

These are based on the ability of complementary nucleic acid strands to come together to form stable double-stranded complexes. The Gen-Probe mycoplasma TC kit (available from Lab Impex in the UK) gives results the same day. Cell culture supernatant or cell extracts are incubated with a ^{3}H-labelled, single-stranded DNA probe. Hydroxyapatite is used to separate hybridized from unhybridized probe before scintillation counting. The disadvantages of this method are cost and the necessity for using a radioisotope.

The Mycotect kit (Life Technologies)

This requires the cocultivation of cells to be tested with 6-methylpurine deoxyriboside. If mycoplasma are present, 6-methylpurine deoxyriboside will be broken down to produce toxic products causing cell death. This is a useful method for those who have no specialized equipment available.

Mycoplasma eradication

Over the years, a wide range of methods have been employed for the attempted eradication of mycoplasma (Gignac, S. M., Brauer, S., Hane, B., Quentmeier, H., Drexter, H. G., 1991; Coronato, S., Vullo, D., Coto, C.E. 1993; LaLinn, M., Bellett, A. J. D., Parsonss, P. G. & Suhrbier, A., 1995). In our laboratory we have had considerable success using MRA (ICN Flow Cat. No. 30–500–44)

Authentication of cell cultures/lines

There is always a risk of inadvertent addition of another cell line to a growing culture. Data have been collected which show (in the USA) that, out of 466 cell lines from 62 different laboratories, 75 were found to be incorrectly identified. A total of 43 cell lines were not of the species expected and 32 lines were incorrect mixtures of 2 or more lines (Ilay, Caputo & Macy, 1992).

To minimize the risk of cellular cross-contamination, only one cell line should be handled in the hood at a time. As stated previously, each cell line should have its own bottle of medium and pipettes should be used for a single manipulation and never replaced into a bottle of medium. Great care must be taken to label each and every cell culture at every stage with designation, passage and date. Areas should be carefully swabbed with alcohol between the handling of each cell line.

The species of origin can be determined by immunological tests, iso-enzymology, cytogenetic analysis and DNA fingerprinting.

Isoenzyme analysis

Isoenzymes are enzymes which catalyse the same reactions but have different electrophoretic mobilities. By determining the mobilities of three iso-enzyme systems (glucose-6-phosphate dehydrogenase, lactic dehydrogenase and nucleoside phosphorylase) using vertical starch gel electrophoresis, it is possible to identify the species of origin of cell lines. A standardized method (the Authentikit system from Innovative Chemistry★) provides for the detection of seven different enzyme reactions. The more enzymes used, the more composite a picture is built up but it is difficult to achieve unique identification of a cell line (see Fig. 8.5).

★Innovative Chemistry, PO Box 90, Marshfield, Massachusetts 02050, USA   Tel: (617)-837-6709. Fax: (617)-834-7325

Karyotyping

It is possible to detect changes in cell cultures and the occurrence of cross-contamination between cell lines by studying the chromosome complement of a cell line. It is however, a complicated technique and requires a highly trained operator of considerable experience. In our laboratory, as an alter-

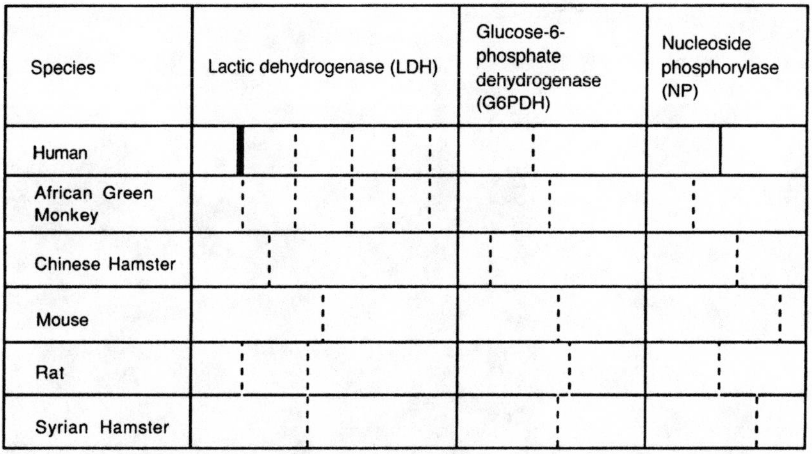

Species	Lactic dehydrogenase (LDH)	Glucose-6-phosphate dehydrogenase (G6PDH)	Nucleoside phosphorylase (NP)
Human			
African Green Monkey			
Chinese Hamster			
Mouse			
Rat			
Syrian Hamster			

Fig. 8.5. Diagram of a vertical starch gel to show an example of the type of pattern that may be obtained for isoenzymes LDH, G6PDH and NP.

native to karyotyping, we use an in-house single locus DNA probe for Southern blots which is specific for human material. (This is suitable for our purposes since the vast majority of our cell lines are of human origin and this technique saves time since the blots can be stripped and reused with the Jeffrey's 33.15 and 33.6 DNA fingerprinting probes. (Cellmark Diagnostics) (Burke et al., 1991).

DNA fingerprinting

A range of probes and methods have been developed using the repetitive sequences found in the animal kingdom. Multilocus probes can identify many loci within a genome, and the use of these probes can detect cross-contamination from a wide range of species. Each cell line has a unique probe fingerprint.

The basis of the technique is to digest genomic DNA from the cells with the restriction enzyme Hinf1 before separating the fragments by agarose gel electrophoresis. The gel is then blotted on to a membrane and hybridized either with ^{32}P-labelled Jeffrey's probes 33.15 and 33.6 or similar probes non-isotopically labelled (from Cellmark Diagnostics)*. Our laboratory favours the latter because of the convenience of not handling radioactivity and because of the number of times the probes are reusable. These probes are sold with a non-radioactive system for detection. Computer-based systems are being developed for the analysis of DNA fingerprints, enabling comparisons to be made between data from many autoradiographs (see Fig. 8.6).

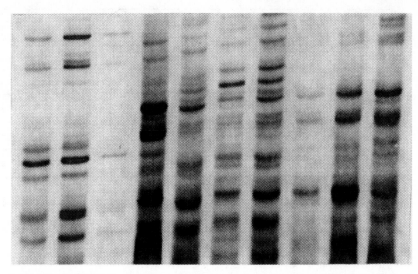

Fig. 8.6. Typical DNA fingerprinting patterns as seen when using Jeffreys' 3.15 probe, as described in the text.

*Cellmark Diagnostic, Blacklands Way, Abingdon Business Park, Abingdon, Oxfordshire OX14 1DY, UK   Tel: 01235 528609   Fax: 01235 528141

The use of PCR in the DNA fingerprinting of cell lines

Amplification of DNA using the polymerase chain reaction (PCR) technique results in a large number of specific sequence DNA copies which can be detected using sequence specific oligonucleotide (SSO) probes. The PCR product is immobilized on a membrane and the SSO probes hybridized, or vice versa (Kirby, 1990)

HLA typing

The histocompatibility system in humans consists of HLA antigens present on the plasma membrane of cells. These antigens provide a polymorphic system for human grouping. Antigens are detected by a complement-dependent cytoxicity test. Whilst this technique can be applied successfully for typing many cell lines, modifications are required in many cases usually due to variations in the form of non-specific antibodies present in either the HLA antisera or the complement.

It is unlikely that all laboratories will be able to set up a range of tests that will cover all aspects of authentication, and it may be useful to note that many of these tests, as well as tests for contamination, are carried out as a service by the cell banks, e.g. ATTC and ECACC.

References

Baltimore, D. (1970). RNA-dependent DNA polymerase in virions of RNA tumour viruses. *Nature (Lond.)*, **226**, 1209–11.

Burke, T. Dolt, G., Jeffreys, A. J. & Wolff, R. (eds). (1991). *DNA Fingerprinting: Approaches and Applications*, pp. 36–370. Basel: Birkhauser Verlag.

Chen, T. R. (1977). *In situ* detection of mycoplasma contamination in cell cultures by Fluorescent Hoechst 33258 stain. *Exp. Cell Res.*, **104**, 255–62.

Coronato, S., Vullo, D. & Coto, C. E. (1991). A simple method to eliminate mycoplasma from cell cultures. *J. Virol. Methods*, **46**, 85–94.

Doyle, A. & Bolton, B. J. (1994). The quality control of cell lines. In *Basic Cell Culture. A Practical Approach*. ed. J. M. Davis, Chap. 8, pp. 243–71. Oxford: IRL Press.

Fogh, J. & Fogh, H. (1964). A method for direct demonstration of pleuro-pneumo-nia-like organisms in cultured cells. *Proc. Soc. Exp. Biol. Med.*, **117**, 899–901.

Gignac, S. M., Brauer, S., Häne, B., Quentmeier, K. & Drexter, H. G. (1993). Elimination of Mycoplasma from infected leukaemia cell lines. *Leukaemia*, **5**(2), 162–5.

Hay, R. J., Caputo, J. & Macy, M. L. (eds). (1992). *ATCC Quality Control Methods for Cell Lines*, 2nd edn, p. 49. American Type Culture Collection.

Hertig, C., Pauli, U., Zanoni, R. & Peterhans, E. (1991). Detection of bovine viral diarrhoea (BVD) virus using the polymerase chain reaction. *Vet. Microbiol.*, **25**, 65.

House, W. & Wadell, A. (1967). Detection of mycoplasma in cell cultures. *J. Path. Bacteriol.*, **93**, 125–32.

Kirby, L. T. (1990). *DNA Fingerprinting*. Chap. 6, p. 115.

LaLinn, M., Bellett, A. J. D., Parsons, P. G. & Suhrbier, A. (1995). Complete removal of mycoplasma from viral preparations using solvent extraction. *J. Virol. Methods*, **52**, 51–4.

Mowles, J. M. & Doyle, A. (1990). Cell culture standards-time for a rethink? *Cytotechnology*, **3**, 107–8.

Peebles, P. T. (1975). An *in vitro* focus-Induction assay for xenotropic murine leukaemia virus, feline leukaemia virus C and the feline–primate viruses RD-114/ccc/m-7. *Virology*, **67**, 288–91.

Rowe, W. P., Pugh, W. E. & Hartley, J. W. (1970). Plaque assay techniques for murine leukaemia viruses. *Virology*, **42**, 1136–9.

Temin, H. M. & Rubin, H. (1958). Characteristics of an assay for Rous sarcoma virus and Rous sarcoma cells in tissue culture. *Virology*, **6**, 669–88.

Temin, H. M. & Mizutani, S. (1970). RNA-dependent DNA polymerase in virions of Rous Sarcoma virus. *Nature (Lond.)*, **226**, 1211–13.

9

Specialized techniques

The main purpose of this book is to provide basic information sufficient for anyone unfamiliar with, or with relatively little experience of, cell culture techniques, to get started and produce comparatively small amounts of cells successfully. In this chapter, an introduction is given to some of the more specialized techniques used in cell culture, including those designed primarily for maximizing the yield of cells or cell product and for immortalizing cells.

Scaling-up

Some techniques are specific to adherent (anchorage-dependent) cells, some to suspension cells, and some can be used for either type.

Specifically for adherent cells

There are various methods available to increase the yield of cells beyond that obtained from the basic culture flask. The main method designed specifically for adherent cells is the use of microcarriers.

Microcarriers

Microcarriers are minute beads made from various materials available from several suppliers in various sizes and specific gravities (see Table 9.1).

Some microcarriers are solid beads allowing cells to grow only on the outer surface and some are 'macroporous' constructed like a sponge where the cells grow not only on the outer surface but within the microcarrier itself greatly increasing the yield (20- to 50-fold). The use of microcarriers enables adherent cells, once attached to the microcarriers, to be grown in a spinner vessel or

Table 9.1. *Suppliers and types of microcarriers*

Supplier	Brand	Type	Material	Diameter (μM)	Specific gravity	Culture type
Asahi Chem. Ind. Co.	Microcarrier	Porous	Cellulose			ST
Biomat Corporation	Informatrix	Porous	Collagen/glycos-aminoglycan	400–800		FB
Hy-Clone Labs.	Cultispher G	Porous	Gelatin	90–150 150–210	1.02/1.03	ST
ICN	Cellagen	Porous	Collagen	100–400		ST
Kirin Co. Ltd	Cellsnow	Porous	Cellulose			ST
Nunc	Biosilon	Solid	Plastic	160–300	1.05	ST
Pharmacia	Cytodex 1,2,3	Solid	Dextran based	131–220	1.03	ST
QDM	Cellfast	Porous				
Schott Glaswerke	Siran	Porous	Glass	400–1000		FB
Schott Glaswerke	Siran	Porous	Glass	4000–5000		IB
SoloHill Engineering		Solid	Plastic	90–150 150–210	1.02/1.04	ST
SoloHill Engineering		Solid	Glass coated	90–150 150–210	1.02/1.04	ST
SoloHill Engineering		Solid	Collagen coated	90–150 150–210	1.02/1.04	ST
Verax Corporation	VX-100	Porous	Collagen	500–600		FB

Note:
Culture types: FB=Fluidized bed culture; IB=Immobilized (fixed) bed culture; ST=Stirred tank culture.

fermenter as a stirred culture allowing comparatively large numbers of cells to be grown in a relatively small volume of medium. It is possible to recover the cells from the microcarriers by the use of a proteolytic enzyme such as trypsin but not all the cells will be recovered if the macroporous type are used. Macroporous microcarriers are ideal where the prime requirement is for extra-cellular product. The method is essentially a simple one and quite within the capabilities of anyone proficient in basic cell culture techniques. Microcarriers, particularly the macroporous type, are increasingly used to grow both adherent and suspension cells in fixed and fluidized bed cultures (see below).

For further reading on microcarriers see Cahn (1990) and Van Wezel (1967, 1977).

Suppliers of microcarriers

HyClone Europe, Dendermondsesteenweg 56, 9300 Aalst, Belgium Tel: +32 53 706090 Fax: +32 53 704171

HyClone Europe Ltd, Nelson Industrial Estate, Cramlington, Northumberland NE23 9BL, UK Tel: 01670 734093 Fax: 01670 732537

HyClone Laboratories Inc., 1725 South HyClone Road. Logan, Utah 84321-6212, USA Tel: (801)-753-4584 Fax: 1-800-533-9450 (USA); 1-801-753-4589 (Other countries)

ICN Biomedicals Ltd, Unit 18, Thame Business Centre, Wenman Road, Thame, Oxfordshire OX9 3XA, UK Tel: 01844 213366 Fax: 01844 213399

ICN Pharmaceuticals Inc., 3300 Hyland Avenue, Costa Mesa, California 92626, USA International. Tel: (714)-545-0113 Fax: (714)-641-7275

Life Technologies Inc., 8400 Helgerman Court, PO Box 6009, Gaithersburg, Maryland 20884-9980, USA Tel: (301)-840-8000 Fax: (301)-670-8539

Life Technologies Ltd, 3 Fountain Drive, Inchinnan Business Park, Paisley PA4 9RF, Scotland, UK Tel: 0141-814 6100 Fax: 0141-814 6317

Nunc A/S, Kamstrupvej 90, PO Box 280, 4000-Roskilde, Denmark Tel: +45 46 35 90 65 Fax: +45 46 35 01 05

Nunc brand: obtainable from Life Technologies Ltd. in the UK

Nalge Nunc International, PO Box 20365, Rochester, New York 14602, USA Tel: (716)-264-3886 Fax: (716)-264-3706

Pharmacia Biotech, 23 Grosvenor Road, St.Albans, Hertfordshire AL1 3AW, UK Tel: 01727 814000 Fax: 01727 814001

Pharmacia Biotech Norden, Djupdalsvagen 20–22, Box 776, 191 27 Sollentuna, Sweden Tel: 08 623 8500 Fax: 08 623 0069

QDM Laboratories, Unit 41C., Derriaghy Industrial Estate, The Cutts, Dunmurry, Belfast BT17 9HZ, UK Tel: Phone and Fax: 01232 301968 'Cellfast'

Sigma Biosciences, Sigma-Aldrich Company, Fancy Road, Poole, Dorset BH12 4QH, UK Tel: 01202 733114 Fax: 01202 715460

SoloHill Laboratories Inc., 4220 Varsity Drive, Ann Arbor, Michigan 48108, USA
Tel: (313)-973-2956   Fax: (313)-973-3029   E-mail: Solohill@ic.net   Web
site: http://WWW.Solohill.com

Techniques common to both suspension and adherent cell culture

Fermenters

Available in sizes from small bench-top units of 1 litre or less in capacity to
large industrial models with capacities of many thousands of litres. Most
manufacturers supply computer-based programs of varying sophistication for
the control of the growth conditions and parameters within the fermenter.
Unless the volumes required are sufficiently large, or the specific conditions
needed demand it, the cost of investment in expensive capital equipment
should be considered very carefully and only pursued if the work cannot be
done by some other method, e.g. spinner flasks.

Recovering cells or supernatant from relatively small volumes of medium
can best be accomplished using a high-capacity low-speed centrifuge
capable of taking up to 6 litres at a time in centrifuge bottles. For larger
volumes, some form of cell separator (elutriator) can be used to avoid
multiple centrifuge runs and reduce handling but they are comparatively
more expensive. If it is essential to maintain sterility when processing large
volumes of suspension to recover either the cells or the supernatant, this is
best accomplished using a system specifically designed for this purpose, e.g.
Sorvall Centritech which gently separates out viable cells, dead cells, cell
debris and clarified liquor. Product loss is marginal and full biological
integrity is maintained. For further reading on this subject see Butler
(1991).

Suppliers of fermenters

Applikon Dependable Instruments bv., PO Box 149, 3100 AC Schiedam, The
Netherlands   Tel: (0)10 462 18 55   Fax: (0)10 437 96 48
Applikon Inc., 1165 Chess Drive, Suite G., Foster City, California 94404, USA
Tel: (415)-578-1396   Fax: (415)-578-8836
B.Braun Biotech International GmbH, Schwarzenberger Weg 73-79, 34212-
Melsungen, Germany   Tel: +49 5661 713704   Fax: +49 5661 713702
Inceltech Groupe, 15 allee de Belle Fontaine, 31100–Toulouse, France   Tel: 33 61
40 85 85   Fax: 33 61 41 51 78
Inceltech UK, Ltd, 22A Horseshoe Park, Pangbourne, Berkshire RG8 7JW, UK Tel:
01734 844888   Fax: 01734 841677

Life Science Laboratories, 15 Ribocon Way, Progress Business Park, Luton, Bedfordshire LU4 9UR, UK   Tel: 01582 597676   Fax: 01582 581495

New Brunswick Scientific Company Inc., Box 4005, 44 Talmadge Road, Edison, New Jersey 08818-4005, USA   Tel: (908)-287-1200   Fax: (908)-287-4222

New Brunswick Scientific (UK) Ltd, Edison House, 163 Dixons Hill Road, North Mymms, Hatfield, Hertfordshire AL9 7JE, UK   Tel: 01707 275733/275707 Fax: 01707 267859

Suppliers of down stream processing equipment

Alfa Laval Separation Ltd, Doman Road, Camberley, Surrey GU15 3DN, UK   Tel: 01276 63383   Fax: 01276 61088.

Du Pont Company, Biotechnology Systems Division, PO Box 80024, Wilmington, Delaware 19880–0024, USA   Tel: (800)-551-2121   Fax: (302)-892-0719

Du Pont (UK) Ltd, Biotechnology Systems Division, Wedgwood Way, Stevenage, Hertfordshire SG1 4QN, UK   Tel: 01438 734680   Fax: 01438 734379

New Brunswick Scientific Inc., 44 Talmadge Road, Edison, New Jersey 08818-4005, USA   Tel: (08)-287-1200   Fax: (908)-287-4222

New Brunswick Scientific (UK) Ltd, Edison House, 163 Dixons Hill Road, North Mymms, Hatfield Hertfordshire AL9 7JE, UK   Tel: 01707 275733/275707 Fax: 01707 267859   'CEPA'

Spinner flasks

Except where extracellular product is the prime requirement, the majority of researchers usually require a healthy homogeneous population of cells in which the majority are at a similar stage of development. To achieve this, the conditions within the culture vessel must be as uniform as possible and the simplest way of providing such conditions is to grow the cells in a stirred vessel. Both suspension cells and adherent cells on microcarriers can be grown in this way. Design of the stirrer paddle and the speed at which it rotates are important. Many designs are available but one which gives complete mixing at the lowest possible speed of rotation should be chosen as this will minimize the damaging effects of paddle shear. Considerable increases in cell numbers per unit volume can be achieved with the use of aeration, either as a sparge (air bubbling through the liquid) or through the semi-permeable walls of silicone rubber tubing. Foam formation at the surface of the liquid needs to be controlled by the use of an antifoam as bubbles may damage the cells (Handa–Corrigan, 1990; Spier & Whiteside, 1990) (see Chapter 2 Equipment for suppliers of stirrer vessels). If one has access to the services of a glassblower, it is quite easy to attach two or more side arms for use with screw caps to 5- or 10-litre (or larger) borosilicate glass bottles, e.g., Pyrex, Schott to produce a comparatively inexpensive stirrer vessel, whilst it is reasonably straight forward to adapt the plastic cap to hold a stirrer shaft.

Roux culture flask

Before disposable plastic bottles became available most adherent cells were cultured on flat-sided prescription bottles of various sizes with aluminium caps and rubber wads (liners). The largest size available was the 500 ml (20 oz) bottle with a surface area of 132 cm^2 per side. To meet the requirement for increased surface area, particularly for vaccine production, specifically designed bottles were manufactured. The smallest, the Roux culture flask (230 cm^2) is still in production but the larger ones (the Pavitski (375 cm^2) and Lepine (1350 cm^2) bottles) have been superseded.

The Roux flask is a borosilicate glass bottle with two flat sides providing approximately 230 cm^2 surface area per side on which to grow anchorage-dependent cells. This bottle can also be used to grow up to 500 ml of suspension cells. The disadvantage of the original design was the requirement for a sterile rubber bung as closure, but they are available with either a screw cap or slip-on stainless steel closure from some manufacturers. However, for the little increase in yield they are somewhat cumbersome and awkward to handle.

Roller bottles

Two sizes of roller bottle are generally available in both disposable plastic and borosilicate glass with 850 cm^2 and 1750 cm^2 surface area, equivalent to approx. 5 × and 10 × the surface area of a large (175 cm^2) flask, respectively. Corning also supply a smaller roller bottle of 490 cm^2. Some manufacturers supply roller bottles with ribbed (pleated) or serrated sides ('extended surface') to increase the available surface area up to as much as 1400 cm^2 in the same sized bottle as the 850 cm^2 and 3500 cm^2 in the same sized bottle as the 1750 cm^2.

Smooth sided roller bottles can also be used to grow suspension cells: 500 ml in the 850 cm^2 size and up to 1200 ml in the 1750 cm^2 size bottle. Several manufacturers supply roller machines (see Chapter 2 under equipment). A less expensive method is to use 2-litre round glass prescription bottles ('burr-lers'), although these are now only available in darkened (amber) glass. However, it is still possible to see adherent (anchorage-dependent) cells growing on the inside wall of these bottles sufficiently well, although it is difficult to make a visual assessment of the pH of the medium.

Fluidized beds/hollow fibres

Although they differ slightly in operation, both these methods are used primarily where the objective is to maximize the production of extracellular product. They provide a large surface area per unit volume and, even with cells that normally grow at low cell densities yields, can be up to 50 times that in an unsupported system. The cells are seeded into a closed system either a bed (substrate) or in hollow fibres.

Hollow Fibres These can be used for both anchorage-dependent (adherent) and suspension cells. The technique is based upon equipment originally developed for haemodialysis using housings (cartridges) containing several thousand minute capillary-like plastic tubules (fibres) with perfusable membrane walls and a molecular weight cut-off (MWCO) of 6,000–10,000 Daltons. Cells are inoculated into the space between the fibres (the extra-capillary space, ECS) where they grow to densities that simulate tissue ($\sim 10^9$ cells per ml). Perfusion of medium through the fibres (the intracapillary space, ICS) provides efficient exchanges of nutrients, gases and metabolites. Microprocessor control of the system is essential. In smaller systems, the exchange of nutrients, metabolites, gases and waste products is accomplished by what is known as the Starling effect. Medium is pumped into the hollow fibres creating relatively high pressure within the ICS forcing medium through the pores of the fibres into the ECS. As medium flows into the ECS, the pressure within the ICS drops progressively along the length of the fibre. Near the far (distal) end of the cartridge the high relative pressure within the ECS forces medium to flow back into the ICS carrying with it waste products from the ECS. However, this effect is inefficient in larger systems causing unequal distribution of nutrients and removal of metabolic waste resulting in the promotion of microenvironments. In larger systems the ECS is connected to an expansion chamber and the ICS to an integration circuit (IC) both of which can be pressurized. This allows the relative pressures within the ECS and the ICS to be counterbalanced creating a pressure gradient similar to but more efficient than the Starling effect. This gradient forces medium from the fibres into the ECS, through the expansion chamber and back to the fibres using valves to create a mono-directional flow. The increased efficiency of this system helps to reduce the development of microenvironments. In some systems the problem of inadequate mixing is overcome by alternating the direction of flow of the medium. Cultures in hollow fibres can be maintained for 6–9 months yielding up to 1 g of product per day from the larger apparatus. Medium can be supplied in sufficient volume from plastic bags of varying sizes. The advantages of hollow fibre cultures over an unsupported system are:

(a) the product is harvested in a more concentrated form;
(b) the volume of medium and serum required is less;
(c) down time is less frequent.

A training course providing both theoretical and practical hands on experience of all aspects of hollow fibre operation is run by the Wolfson Cytotechnology Laboratory at Surrey University.

For further reading see Knight (1989), Looby & Griffiths (1990), Van Wezel (1977) and the Pharmacia microcarrier handbook, *Microcarrier Cell Culture: Principles and Methods* (1981).

Suppliers of hollow fibre culture apparatus

Cellex Biosciences Inc., 8500 Evergreen Boulevard, Coon Rapids, Minneapolis, Minnesota 55433, USA Tel: (612)-786-0302 Fax: (612)-786-0915 ('Acusyst', 'Endotronics')

New Brunswick Scientific, Edison House, 163 Dixons Hill Road. North Mymms, Hatfield, Hertfordshire AL9 7JE, UK Tel: 01707 275733 Fax: 01707 267859 ('Celltronics')

New Brunswick Scientific Inc., 44 Talmadge Road, Edison, New Jersey 08818-4005, USA Tel: (908)-287-1200 Fax: (908)-287-4222 ('Celltronics')

Tecnorama AG, Industriestrasse 44, CH-8304 Wallisellen, Switzerland Tel: (01) 8 30 22 77 Fax: (01) 8 30 78 52 ('Tecnomouse')

Fixed bed (Packed bed/immobilized bed)/fluidized bed. In this system, the substrate in and/or on which the cells grow is not stirred but perfused. Both fluidized and fixed bed cultures are set up and seeded in the same way but, after 48 h or so, the modes of operation become different. Fixed bed apparatuses typically have a vertical cylindrical body with an aspect ratio (height to width) of 3:1. Medium is circulated via a reservoir in which the culture conditions are monitored and controlled. Fluidized beds have a higher aspect ratio, around 8:1 and medium circulation is achieved either by pumping or air circulation.

These types of culture have similar advantages and disadvantages to those for hollow fibres and are capable of long-term continuous operation. For a general review of the principles and procedures, see Griffiths (1990).

Apart from several types of microcarrier (see Table 9.1), other materials have been tried as the bed matrix, e.g. stainless steel wool and springs, cellulose fibres and cellophane: those substrates that are currently available commercially include the following:

Ceramic A form of kaolinite which are hollow porous microspheres available in two sizes: 20–50 μM with 10 μM cavities and 50–75 μM with 20 μM cavities, giving respective surface areas of 5 m^2/g^{-1} and 9.5 $m^2/g.^{-1}$. They can be used in stirred tanks but because of their relatively high specific gravity and shape are abrasive and cause cell damage. For further reading, see Berg & Bodeker, (1988).

Suppliers of ceramic matrices

Cellex Biosciences Inc., 8500 Evergreen Boulevard, Coon Rapids, Minneapolis, Minnesota 55433, USA Tel: (612)-786-0302 Fax: (612)-786-0915 ('Opticell')

ECC International Ltd, Household Products Division, John Keay House, St Austell, Cornwall PL25 4DJ, UK Tel: 01726 74482 Fax: 01726 623019 ('Biofix')

Polymer sponge matrices Two of the types available are a non-woven polyester fabric, e.g. FibraCel and polystyrene twisted ribbons (e.g. Heli–Cel). They need surface treatment before use. For further reading, see Leighton *et al.* (1967)

Suppliers of polymer sponge matrices

Bibby Sterilin Ltd, Tilling Drive, Stone, Staffordshire ST15 0SA, UK Tel: 01785 812121 Fax: 01785 813748

Porous silicone rubber, e.g. ImmobaSil This material is inert, non-toxic, biocompatible, mechanically robust, soft with a high surface area ($\sim$50 000 m^2/m^3) and with good electostatic properties: autoclavable, re-usable with low levels of free cells. It will support the growth of $\sim$6.75 × 10^{7}cells per ml of carrier for adherent cells and $\sim$2.09 × 10^7 cells per ml of carrier for suspension cells. Average yield is claimed to be 18.5 μg/10^6 cells/day. Available in 1 mm × 1mm blocks or as a sheet. The sheet can be used to coat the inner surface of tubing and allows good perfusion of medium, metabolites and gases.

Suppliers of silicone rubber matrices ('ImmobaSil')

Ashby Scientific Ltd, Unit 11, Atlas Court, Coalville, Leicestershire LE67 3FL, UK
Tel: 01530 832590   Fax: 01530 832591

Permeable culture supports

Polarized primary epithelial cells are often cultured on permeable supports. These enable cells to repolarize in culture and be fed from below. This is particularly useful for airway epithelium, for example, where an air/cell interface can be maintained on the epithelial surface in culture with a cell/liquid interface below the cells provided by the permeable support. Permeable cell culture supports are available from a number of companies, e.g. Millipore, Falcon, Becton Dickinson, see Chapter 2.

Coated surfaces

Many cell types adhere better to plastic surfaces if these have been precoated with an extracellular matrix component such as collagen. In the volume on *Epithelial cell culture*, detailed methods for collagen–coating substrates are presented. In addition, precoated substrates may be purchased from a number of cell culture plastics suppliers, e.g. Falcon, Becton Dickinson, Millipore, see Chapter 2.

Immortalization of cells (transformation)

Immortalization, or the ability of a cell to multiply indefinitely, is a multistep process of irreversible genetic change that, for some cell species, can and will occur naturally but the process takes time and is not always successful. This natural process of genetic change leading to the immortalization of a cell line commences at the time of the first passage but the first obvious evidence occurs if the cell strain survives the one, or occasionally, two phases referred to as 'crises' during the early passage stages in its culture. It is not until around, or after, the tenth passage that the cells will start to grow continuously again. This can be a fairly time-consuming process particularly as the crises are periods in which the cells apparently stop growing, sometimes for many days or even weeks. Whilst the cells are in the period of crisis the medium should be changed weekly. The process is termed transformation but, unlike bacterial transformation, the process of phenotypic modification of mammalian cells may not involve the uptake of new genetic material.

Another crisis period is encountered around the 45 to 55 passage level and many cell lines, particularly diploid cell lines, die out at this point. Those that survive will continue to propagate continuously without any further crises and are then termed 'immortalized', i.e. they have an infinite lifespan. One of the characteristics of the process of transformation is a change in the phenotype, which is usually much more evident in adherent cells than in suspension cells. The other obvious changes that occur are an increased growth rate, a higher plating efficiency and occasionally they become tumourigenic. Greater success in transforming cells can be obtained by using a virus or by exposure to a chemical carcinogen to bring about the phenotypic modification. Viral immortalization involves the use of a viral agent such as polyoma virus or Epstein–Barr virus (EBV) often used to transform lymphocytes and Rous Sarcoma virus used to transform normal avian cells. Viral transformation also brings about the genotypic and phenotypic changes characteristic of natural transformation. These procedures need to be conducted under controlled environment conditions in an area specifically designated for the purpose separate from other working areas and with restricted entry.

Cloning

It is too often assumed that a cell line will continue to grow passage after passage without any significant alteration to its karyotype and, in the case of cells with more specialized properties, any loss of the specific property. However, preservation of specific cell lines and their specialized properties are major problems in cell culture, due to genetic drift and the selective overgrowth of unspecialized cells. In several fields of research the ability to clone cells is a necessary requirement and, although comparatively simple to perform, is labour intensive and tedious.

There are three main methods of cloning cells, the soft agar technique for suspension cells, the use of cloning rings for adherent cells and the limiting dilution technique which is useful for both types of cell. More detailed descriptions of these methods can be found in specialist texts.

Transfection

This the process in which DNA, usually in the form of a recombinant plasmid, is introduced into a host cell. In the field of molecular biology, there are many applications for cell transfection which fall into two main categories: short-term or transient transfections and long-term or stable transfections. In the former case, cell lysates are assayed 48–72 hours post-transfection. In the latter case, the DNA used for the transfection should include an expression cassette for a drug-selectable marker (for example, neomycin resistance); the drug is then added to the culture medium 24–48 hours post-transfection resulting in death of most of the cells. However, those cells which stably integrate the transfected DNA into their genomes will survive to produce colonies of cells, which can be separately amplified and characterized. The efficiency of the DNA uptake varies enormously between cell lines and consequently a number of different delivery methods have been developed. The particular method selected will therefore depend not only on the cell line being used but also on the type of experiment envisaged. The most frequently used methods are as follows.

Calcium phosphate

A precipitate is formed containing DNA and calcium phosphate which is taken up by the cells. This method is economical and applicable to a wide variety of cells, but many epithelial lines are inefficiently transfected. In addition, lines selected from long-term transfections have frequently integrated many copies of the transfected DNA as long concatamers and this may not be desirable.

DEAE-dextran

DNA bound to the polymer DEAE-dextran is taken up by the cells and, again, this method is economical and is useful for short-term transfections. However, it is not recommended for stable transfections.

'Lipofection'

Cationic liposomes surround the DNA and carry it into the cell by fusing with the membrane. This process is very efficient for both types of transfection, but the different lipid preparations are not equally effective on all cells and it can be expensive to determine which lipid is best for your cells. Some companies now market 'tester kits' with a number of different preparations to assay.

Electroporation

A brief high voltage electric current is applied to the cells to punch pores in the membrane through which the DNA can enter. This is the best way to

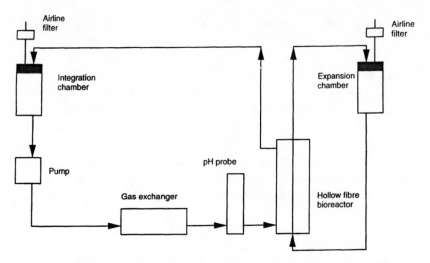

Fig. 9.1. Schematic of a hollow fibre apparatus (based on the Acusyt system). See text for explanation.

transfect non–adherent cells and it can be very useful for long–term transfections in all cells when only one or two copies of transfected DNA are wanted per cell. As relatively few copies of DNA enter each cell, it is not always useful for short term transfections. The main drawback is that the exact conditions need to be determined empirically for each cell type and the electroporator and disposable cuvettes are expensive.

Apart from the different techniques themselves, there are a number of treatments that can be applied to the cells after transfection to try and improve either DNA uptake, e.g. glycerol shock or the half-life of the DNA once inside the cell, e.g. the lysosome inhibitor, chloroquine. For further reading, see Hames & Higgins (1993) and Latchman (1993).

Hybridomas

Hybridomas are heterokaryons resulting from the fusion of myeloma (plasmacytoma) cells with antibody-forming rodent spleen cells, a technique developed by Kohler and Milstein in 1975 to produce monoclonal antibodies. It should be noted that the technique involves a large amount of work and demands a great deal of commitment and should therefore not be considered if a polyclonal antibody would be adequate.

Most polyclonal antisera contain a large number of antibodies exhibiting a wide range of affinities and extreme heterogeneity. On the other hand, monoclonal antibodies exhibit a well-characterized affinity and homogeneity.

A full description of the principles and practices involved are beyond the scope of this volume, but there are several excellent books on the subject including those by Goding (1986), and Harlow & Lane (1988). The following (illustrated in Fig. 9.2) is but a brief description of the principles.

A rodent (usually a mouse) is inoculated with the particular antigen against which it is hoped to raise an antibody and the animal monitored for antibody production in the usual way. It is important to have a culture of mouse myeloma cells lacking the enzyme HGPRT (hypoxanthine guanine phosphoribosyl transferase) growing at the time the mouse is ready to process, e.g. NSO, P3U1 obtainable from ECACC or ATCC.

The spleen is removed from the mouse and the tissue disaggregated and washed. The prepared cells are mixed with a suspension of myeloma cells in the presence of either a virus, e.g. Sendai, EBV, or a high concentration of polyethylene glycol (PEG). Membrane fusion will occur to form multinucleate cells called heterokaryons, which are usually genetically unstable. The cell suspension will be a mixture of unfused myeloma cells, unfused spleen cells and heterokaryons (hybridomas). To separate these out the suspension is diluted and cloned in the presence of a selective medium (HAT) developed by Littlefield (1964) containing hypoxanthine, aminopterin and thymidine. In this medium the main biosynthetic pathway for guanosine is blocked by the folic acid antagonist aminopterin. In normal cells there is an alternative 'salvage' pathway in which the nucleotide metabolites hypoxanthine or guanine are converted to guanosine monophosphate via the enzyme HGPRT. Any unfused myeloma cells lack the enzyme and cannot use the salvage pathway and will die. Any unfused spleen cells will also not grow. Only fused cells will grow because the spleen cells provide the enzyme for the immortal myeloma cells to utilize. The cloning process is long and tedious before a clone or clones producing the desired antibody is isolated. Because of chromosome loss the cells should be cloned a minimum of twice to isolate true antibody producing cells.

Insect cell culture

The culture of insect cells has become increasingly important in many areas of the biological sciences during the last decade or so including biochemistry, endocrinology, genetics, molecular biology, physiology and virology, particularly for use with the Baculovirus system for over-producing recombinant proteins in eukaryotic cells. The development of insect cell culture

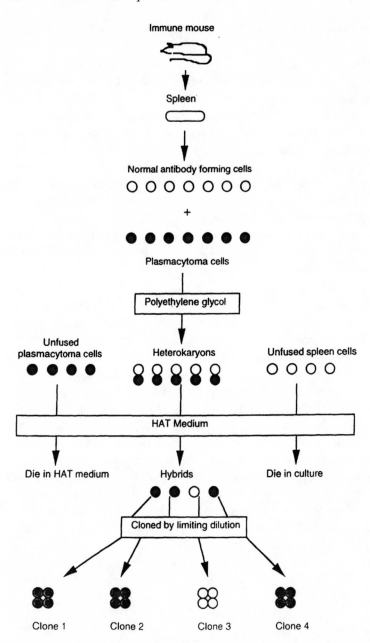

Fig. 9.2. Diagram of the basic steps involved in producing monoclonal antibodies.

has its origins with the work of Goldschmidt who cultured explants of the Cecropia silkmoth *Hyalophora cecropia* during the early part of this century. The success of this, and subsequent attempts, were limited by the lack of an adequate medium to support cell growth and survival. It was not until the late 1930s that any major advances in insect cell culture occurred through the work of Trager, who made systematic attempts to determine the nutritional requirements for insect cells based on the physical and chemical properties of the insect haemolymph. His work proved it was possible for insect cells to survive and remain physiologically active *in vitro*, even if it were only for a matter of a few days. Since the 1950s, major strides have been made to simplify and improve media formulation and preparation and simplify the cell culture techniques. This simplification has resulted in the exponential growth of the use of insect cell culture to the point where today there are over 200 cell lines from more than 75 species and eight orders of arthropods available from major cell repositories. Most commercial media manufacturers supply a range of insect cell culture media. For further detailed information, see the work of Fraser (1989), Granados & Hashomoto (1989), Hink & Branson (1985), Maeda (1989), Payne (1982) and Reddy (1977).

References

Berg, G.J. & Bodeker, B.G.D. (1988). Employing a ceramic matrix for the immobilisation of mammalian cells in culture. In *Animal Cell Biotechnology*. ed. R. E. Spier & J. B. Griffiths, vol.3, pp.321–35. London.: Academic Press.

Butler, M. (ed.) (1991). *Mammalian Cell Biotechnology*, Oxford.: IRL Press.

Cahn, F. (1990). Biomaterials aspects of porous microcarriers for animal cell culture. *Trends Biotechnol.*, **8**, 131–6.

Fraser, M. J. (1989). Expression of eukaryotic genes in insect cell cultures. *In-Vitro*, **25**, 225–35

Freshney, R. I. (1994). *Culture of Animal Cells,* 3rd edn. New York: Wiley-Liss.

Goding, J. W. (1986). *Monoclonal Antibodies: Principles and Practice*, 2nd edn. London: Academic Press.

Granados, R. R. & Hashomoto, Y. (1989). Infectivity of baculovirus to cultured cells. In *Invertebrate Cell System Applications*, ed. J. Mitsuhashi, Boca Raton, Florida, USA: CRC Press,

Griffiths, B. (1990). Advances in animal cell immobilisation technology. In *Animal Cell Biotechnolgy*, eds. R. E. Spier & J. B. Griffiths, vol. 4, pp. 149-66. New York: Academic Press.

Hames, B. D. & Higgins, S. J. (eds) (1993). *Gene Transcription*, pp. 87–120. Oxford: IRL Press.

Handa-Corrigan, A, (1990). Oxygenating animal cell cultures: the remaining problems. In *Animal Cell Biotechnology*, eds. R. E. Spier & J. B. Griffiths, pp. 123-32. London: Academic Press.

Harlow, E. & Lane, D. (1988). *Antibodies—A Laboratory Manual*. Cold Spring Harbor Laboratory Press, Cold Spring Harbor, New York.

Hink, W. F. & Bezanson, D. R. (1985). Invertebrate cell culture media and cell lines. In *Techniques in Setting Up and Maintenance of Tissue and Cell Culture*, ed. E. Kurstak. Ireland: Elsevier Scientific Publishers.

Kadouri, A. & Zipori, D. (1989). Production of anti-leukaemic factor from stroma cells in a stationary bed reactor on a new cell support. In *Advances in Animal Cell Biology and Technology for Bioprocesses*, pp. 327–30, eds. R. E. Spier, J. B. Griffiths, J. Stephenne & P. J. Crooy. Tiptree, Essex, UK: Courier International.

Knight, P. (1989). Hollow fiber bioreactors for mammalian cell culture. *Biotechnology*, **7**, 459–61.

Kohler, G. & Milstein, C. (1975). Continuous cultures of fused cells secreting antibody of predicted specificity. *Nature*, **246**, 495–7.

Latchman, D. S. (ed.) (1993). *Transcription Factors*, pp. 148–52. Oxford : IRL Press.

Leighton, J., Justh, G., Esper, M. & Kronenthal, R. L. (1967). Collagen-coated cellulose sponge: three dimensional matrix for tissue culture of walker Tumour 256. *Science*, **155**, 1259–61.

Littlefied, J. W. (1964). Selection of hybrids from matings of fibroblasts *in-vitro* and their presumed recombinants. *Science*, **145**, 709–10.

Looby, D. & Griffiths, B. (1988). Fixed bed porous glass sphere (porosphere) bioreactors for animal cells. *Cytotechnology*, **1**, 339–46.

Looby, D. & Griffiths, J.B. (1990). Immobilisation of animal cells in porous carrier culture. *Trends Biotechnol.*, **8**, 204–9.

Maeda, S. (1989). Expression of foreign genes in insects using Baculovirus vectors. *Ann. Re. Entomol.*, **34**, 351–72.

Microcarrier Cell Culture: Principles and Methods (1981). Uppsala, Sweden: Pharmacia Biotech.

Payne, C.C. (1982). Insect viruses as control agents. *Parasitology*, **84**, 35-77.

Reddy, D. V. R. (1977). Techniques in invertebrate tissue culture for the study of plant viruses. *Methods Virol.*, **6**, 393–434.

Spier, R. & Whiteside, P. (1990). The oxygenation of animal cell cultures by bubbles. In *Animal Cell Biotechnology*, eds. R. E. Spier & J. B. Griffiths, pp. 133–48. New York: Academic Press Ltd.

Van Wezel, A. L. (1967). Growth of cell strains and primary cells on microcarriers in homogeneous culture. *Nature (Lond.)*, **216**, 64–5.

Van Wezel, A. L. (1977). The large scale cultivation of diploid cell strains in microcarrier culture. Improvement of microcarrier. *Dev. Biol. Standard*, **37**, 143–7.

Wang, G., Zhang, W., Freedman, D., Jacklin, C., Eppstein, L. & Kadouri, A. (1992). Modified CelliGen packed bed bioreactors for hybridoma cell cultures. *Cytotechnology*, **9**, 41–9.

10

Health and safety

Safety is primarily the application of common sense, care and caution. Long before any legislation was introduced, it was fully appreciated that laboratories as workplaces had certain inherent dangers associated with them and, as a result, most persons working in them took care not to endanger themselves. This also meant that other members of staff were comparatively safe and is the main reason why laboratories have always been statistically one of the safest places in which to work. Even today, most laboratories still operate with a philosophy of safety regardless of any introduced legislation.

Instruction in practical safety relating to a specific laboratory should be available from a senior member of the scientific or technical staff and/or from an organizational safety officer where there is one.

The following are a few general guidelines that should be followed wherever practicable when working in a laboratory:

1 Always wear a laboratory coat (preferably of the newer side-fastening 'Howie' type if possible).
2 Always wear disposable gloves when working. Both coats and gloves should be removed when leaving the laboratory.
3 Never eat, drink, chew, store food, smoke or apply make-up in the laboratory.
4 Never use the mouth to fill or discharge pipettes. Always use some form of manual pipette filler.
5 Hands must be disinfected or washed immediately if contamination is suspected, after handling viable materials and also before leaving the laboratory. Flush eyes and or mucous membrane areas immediately with water if they have come in contact with blood or other viable materials and seek medical assistance.
6 Perform all procedures so as to minimize spillages and the production of aerosols.

7 Avoid the use of 'sharps' except where there is no alternative. Accidents with sharps (needles, scalpels and broken glass) account for the largest proportion of laboratory accidents. Exercise great care when disposing of used hypodermic needles, particularly when resheathing them for disposal. *Never* bend used needles. Some laboratories discard both the needle and syringe together into a 'sharps' disposal container to avoid contact with the needle.

8 Have available the appropriate disinfectant(s) to deal with any microbiological or cellular spillage (see further).

9 Have available decontaminants for chemical, toxic or radioactive spillages where appropriate.

10 Always wash down bench-tops with an appropriate disinfectant after use. Always check any reusable containers on a regular basis and decontaminate as required and as appropriate.

11 Specimens of potentially infectious material should be transported in a closed container, which prevents leakage during collection, processing and storage.

12 Use a tissue culture cabinet or biological safety cabinet for the manipulation of tissues.

13 Dispose of all waste in the correct manner (see further).

14 Secure laboratory gas cylinders to a wall or bench. If they fall, they can cause bruising, lacerations or even broken bones. Should the reduction gauge be broken off a falling cylinder, it has been known for the rapid release of gas to propel the cylinder at great speed across the laboratory.

15 Take great care when handling liquid nitrogen. Always wear insulated gloves and a face visor.

16 Have available bottles of sterile eye wash displayed in a prominent position. Ensure that all staff are familiar with the location and method of operation in an emergency.

17 Ensure that all staff are familiar with the location and method of operation of all types of fire extinguisher and the fire alarms found in the laboratory.

Legislation

In the UK, the introduction of the Health and Safety at Work Act (1974) (HASAWA) brought laboratories as work places, and those employed in them, under government legislation for the first time. Since then, the scope of the Act has been widened by the introduction of continually expanding legislation to cover the more specialized areas of work increasingly found in biological laboratories. This now includes legislation introduced by the EEC and implemented by the UK government.

For anyone involved in cell culture in the UK, the following legislation is relevant, and it is important for all to be familiar with the requirements laid out in them.

Health and Safety at Work, etc. Act (1974)

This statute has two main thrusts: the responsibilities of the employer for the workforce (Part 1, Sections 2–5 and 9) and the responsibilities of the employee for themselves and those with whom they work (Part 1, Sections 7 and 8). The Act also describes the role and powers of the Health and Safety Executive (HSE), the role of the Employment Medical Advisory Service (EMAS), and contains amendments to other existing acts relating to buildings and laboratories. The Act has been amended by the addition of two further pieces of legislation: The Management of Health and Safety at Work Regulations (1992) effective from the 1st. January 1993 implementing EEC Directives 89/391/EEC (*The Framework Directive*) and 91/383/EEC (*The Temporary Workers Directive*); and the Management of Health and Safety at Work (Amendment) Regulations (1994) effective from 1st December 1994 implementing Articles 4 and 7 of EEC Directive 92/85/EEC dealing with pregnancy and the workplace.

Control of substances hazardous to health regulations (1988) (COSHH)

These are a set of statutory regulations made under the HASAWA (1974): they also describe a management system enabling employers to prevent or control exposures to hazardous substances in the workplace with the object of reducing ill health. A substance (or preparation) is defined as being 'hazardous to health' if it:

- is listed in Part 1 of the Approved Supply List as being dangerous for supply under CHIP (Chemicals (Hazard Information and Packaging for Supply) Regulations 1994) SI 1994 No. 3247);
- is listed in Schedule 1 of COSHH as having a 'maximum exposure level' or has an 'occupational exposure standard';
- is a biological agent being any micro-organism, cell culture, or human endoparasite, including any which have been genetically modified, which may cause any infection, allergy, toxicity or otherwise create a hazard to human health;

- is dust of any kind in substantial concentration;
- is any substance 'of comparable hazard.'

It is the duty of every management in relation to such substances when used for work purposes to take the following steps:

Step 1: To assess the risk
Step 2: To provide controls
Step 3: To use controls
Step 4: To maintain controls
Step 5: To inform, instruct and train
(Step 6: To monitor exposure)
(Step 7: To provide health surveillance)

All these points have to be followed by anyone involved in culturing animal cells. The 1988 Act was amended twice, in 1991 and 1992, and then all were combined in an update of the Act in 1994. This introduced new and revised maximum exposure limits for certain substances and new provisions to implement the EC Biological Agents Directive as Schedule 9. It revoked the COSHH 1988 Regulations, the two amendments and The Health and Safety (Dangerous Pathogens) Regulations 1981. Importantly, it defined 'biological agent' and revised the definition of 'micro-organism' so that, in future, all reference will be to 'biological agent' and not to 'micro-organism'. They are now classified into four groups on the basis of their infection risk. Most, if not all, cell lines will come under Group 1: agents unlikely to cause human disease. Because of the uncertain risk associated with some malignant cells, they may be considered to represent a potentially higher risk than covered by just Group 1 and theoretically at least should be considered to come under Group 2: agents that can cause human disease and may be a hazard to employees; it is unlikely to spread to the community and there is usually prophylaxis or treatment available. For this reason, all cell cultures should be handled in at least a Class 2 biological safety cabinet or tissue culture cabinet.

Reporting of injuries, diseases and dangerous occurrences regulations (1985) (RIDDOR) (SI 1985 No. 2023)

This 'imposes a duty on the responsible person (normally the employer) to report to the Health and Safety Executive (HSE) (or, where appropriate, other enforcing body such as the local authority) certain classes of accident, disease and dangerous occurrence. At the same time, the Regulations lay down what records must be kept.'

An outline of US regulations

In the United States, basic guidelines for workers dealing with biological materials are established in the Occupational Safety and Health Standards, part 1910, effective in March 1992. Section 1910.1030 (bloodborne pathogens) defines standards for government, employers and employees.

Institutions are required to appoint safety officers to implement the guidelines, and departments/divisions are encouraged to appoint their own officers who are familiar with particular problems in that unit. Each laboratory is asked to establish operational safety protocols to be approved by the safety officers and kept available in laboratories for reference on a daily basis.

Employers are responsible for providing safety supplies, environmental controls, Hepatitis B vaccinations and training programmes sanctioned by the US Occupational Safety and Health Administration (OSHA) for employees at no cost to the employees. Employees are responsible for using the safety equipment provided and for following the practical guidelines outlined, which includes practices for housekeeping/janitorial staff as well as laboratory professionals.

Specific guidelines are given for the characteristics of equipment and supplies, symbols (biohazard symbol), signs and labels. Practical methods for dealing with laboratory hazards of all sorts are specified (which agree with European practice as detailed in this chapter). Requirements for the keeping of training records, exposure records, vaccination records, and confidential medical records are discussed.

In cases where exposure to biologically hazardous material has occurred, the guidelines for reporting incidents, medical monitoring and follow-up are outlined. The United States Environmental Protection Agency has an established registry of disinfectants, requiring submission of efficacy data from bacterial, fungal, mycobacterial and viral testing before disinfectant labelling is approved.

Disposal of waste

Sharps

Hypodermic needles, scalpels, etc. must be placed in a rigid container designed specifically for the purpose. Never fill these containers more than two-thirds full. Broken glass should be disposed of into tough cardboard boxes specifically designed and labelled for the purpose.

Glassware

Should be immersed in a dilute solution of an appropriate disinfectant, e.g. 2% Chloros (Sodium hypochlorite). Leave overnight before cleaning.

Cultures

Should be graded as clinical waste and be discarded into bags designed for autoclaving.

Spillages

Breakage of containers or spillage of cultures, whether in a laboratory shaker, centrifuge or on the laboratory floor, will usually produce a large amount of aerosols which may be potentially hazardous. If such a spillage occurs, follow these guidelines to minimize exposure.

1 Do not bend over the breakage/spillage to inspect the damage, as this may expose you to a concentrated aerosol.
2 If the spill is within the laboratory, cover it with paper towels and gently pour on some disinfectant working from the outside in. Alternatively, cover with 'Klorsept' granules. Leave for 30 min. and then transfer to the clinical waste.
3 If the spill is within a shaker or centrifuge do not open it. Shut off the power and leave for 30 min. This should be long enough for the aerosol to settle. Open the lid carefully in order to avoid creating further aerosols. Always use a non-corrosive disinfectant, e.g. Cidex, Virkon.
4 If the spillage is sufficiently large, evacuate the room, close the door and inform the Safety Officer.

Disinfectants

Never try to retrieve anything discarded into disinfectant. Too much reliance is often placed in the efficiency and efficacy of disinfectant solutions with the erroneous assumption that everything in it will be dead and the solution 'safe'. This is, in fact, too often not the case. Always wear disposable gloves when handling neat disinfectant.

Sodium hypochlorite, e.g. Chloros (UK); Chlorox or Purex (USA)

Action

Disinfects by active oxidation by chlorine-free radicals.

Range

Effective against bacteria, viruses (including Hepatitis B), fungal and bacterial spores.

Uses

2% solution for laboratory discard jars. 10% for swabbing small spills.

Supplier
Hays Chemical Distribution, 215 Tunnel Avenue, London SE10, UK
Tel: 0181-293 4343

Klorsept

For larger spillage control.

Comments

Recommended for liquid waste, blood and serum. Rapidly inactivated by organic material – exhausted (inactive) when indicator changes from purple to yellow/green.

Very corrosive to metals and skin, etc. Wear protective clothing and visor when handling the neat solution. Can give off chlorine gas as it oxidizes.

Supplier
Medentech Ltd. Whitemill Industrial Estate, Whitemill Road, Wexford, Ireland
Tel: 003535341809   Fax: 003535341271

Peroxygen, e.g. Virkon

Action

Disinfects by oxidation.

Range

Effective against bacteria, all 18 families of mammalian viruses, yeasts and fungal and bacterial spores.

Uses

Contains surfactants. Recommended for cleaning incubators and centrifuges. 1% solution for swabbing small spills. Maximum solubility is 3%. Dry granules for larger spillage control.

Comments

Recommended for fluids, blood and serum. Non-corrosive, non-bleaching and biodegradable. Decomposes above 70 °C with the evolution of sulphur dioxide.

Manufacturer

Antec International Ltd, Chilton Industrial Estate, Windham Road, Sudbury, Suffolk CO10 6XD, UK   Tel: 01787 377305   Fax: 01787 310846

Phenolics, e.g. Wescodyne also known as Tegodyne or Tegodor (UK and USA); Amphyl, Vesphene (USA)

Action

Disinfects by the precipitation of proteins.

Range

Effective against some bacteria and viruses (not Hepatitis B).

Uses

0.5% solution for hand washing. 2% for glassware and instruments.

Comments

Not recommended for discard jars, or use with blood and serum.

Supplier

Synchem (UK) Ltd, PO Box 170, Staines, Middlesex TW18 4UZ, UK   Tel: 01784 449966.   Fax: 01784 466207.

Surface-active agents, e.g. Roccal (Benzalkonium chloride) (UK and USA); BacDown (USA)

Supplied as a 10 × concentrate.

Action

Cationic and amphoteric detergents.

Range

Very limited number of bacteria and viruses. Effective as a fungicide.

Uses

1% solution for incubators and water baths.

Comments

Bacteriostatic not bacteriocidal at low concentrations. Inactivated by organic matter and soap. Can be used on the skin without ill effect. Mildly corrosive to stainless steel on prolonged soaking, hence not recommended by incubator manufacturers.

Supplier

Sanofi Winthrop Medicare, One Onslow Street, Guildford, Surrey GU1 4YS, UK

Gluteraldehydes, e.g. Cidex

Action

Aldehyde reacts with amino groups of proteins.

Range

Effective against bacteria, viruses (including Hepatitis B) fungal and bacterial spores.

Uses

2% solution can be used to swab shakers and centrifuges.

Comments

Action is relatively slow. Leave for at least 1 hour after application. Irritant to eyes, skin and respiratory mucosa. May cause dermatitis. Non-corrosive to metals.

Supplier
Southern Syringe Services Ltd, 303 Chase Road, London N14, UK   Tel: 0181-882 1971

Alcohol, e.g. ethanol, methylated spirits (IMS), isopropanol

Action

Precipitation of proteins.

Range

Some bacteria and viruses. Not effective against spores.

Uses

As a 70–85% solution for bench-tops, cabinets and hard surfaces.

Comments

Volatile and flammable. Use care in enclosed areas. Neat alcohol is relatively ineffective, so always dilute. Not suitable for the inactivation of viral and plasmid DNA. Can be used on the skin without any ill-effect.

Bibliography

UK Regulations

Health and Safety at Work, etc. Act (1974). Chap. 37. 1986 reprint with corrected errors. ISBN 0 10 543774 3.

Management of Health and Safety at Work Regulations (1992) SI No. 2051.

Management of Health and Safety at Work (Amendment) Regulations (1994) SI No. 2865.

Control of Substances Hazardous to Health Regulations (1994). SI No. 3246.

Control of Substances Hazardous to Health Regulations (Northern Ireland) (1990).

COSHH and Peripatetic Workers HS(G) 77.

Health Surveillance Under COSHH: Guidance for Employers.

A Step-by-Step Guide to COSHH Assessment HS(G) 97.

Approved Code of Practice: Biological Agents.

Occupational Exposure Limits 1994. EH40.

Maintenance, Examination and Testing of Local Exhaust Ventilation HS(G) 54.

Advisory Committee on Dangerous Pathogens: Categorization of Biological Agents. According to Hazard and Categories of Containment. 3rd edn 1994.

Health Services Advisory Committee: Safe Disposal of Clinical Waste.

Reporting of Injuries, Diseases and Dangerous Occurrences Regulations (RIDDOR) (1985). ISBN 0 7176 0432 2

First Aid at Work: Health and Safety (First Aid) Regulations (1981). ISBN 0 7176 0426 8

Electricity at Work Regulations (1989).

Safety Representatives and Safety Committees.

Noise at Work.

Manual Handling Operations Regulations.

Guidance on Provision and Use of Work Equipment Regulations.

Workplace Health, Safety and Welfare.

EEC Regulations

EEC: Council Directive 89/391/EEC (The Framework Directive) OJ No. L183 (29 June 1989).

EEC: Council Directive 91/383/EEC (The Temporary Workers Directive) OJ No. L206 (29 July 1991).

EEC: Council Directive 92/85/EEC (Pregnancy and the Workplace) OJ No. L348 (28 November 1992).

EEC: Council Directive 90/679/EEC (Exposure to Biological Agents) OJ No. L374 (31 December 1990).

Personal Protective Equipment.

International legislation

Australia
National Occupational Health and Safety Commission Act 1985.
Occupational Safety and Health Act 1984.
Occupational Health and Safety (Commonwealth Employment) Act 1991.
Occupational Safety and Health Regulations 1988.
Occupational Safety and Health Act 1994 (Malaysia).

Canada
Occupational Health and Safety Act (OHSA).
Control of Substances Hazardous to Health Regulations 1994.

USA
Occupational Safety and Health Act (1970). Amended 1990.
Occupational Safety and Health Standards Part 1910.
OSHA Technical Manual.
Hazardous Waste and Emergency Response.
Exposure to Hazardous Chemicals in Laboratories.
Occupational Exposure to Bloodborne Pathogens.

Organizations

Australia
Australian Institute of Health and Welfare, 6A Traeger Court, Fern Hill Park, Bruce
    ACT 3617, Australia   Tel: +61 6244 1000   Fax: +61. 6244 1299

Canada
The Canadian Centre for Occupational Health and Safety, 250 Main Street East,
    Hamilton, Ontario L8N 1H6   Tel: (905)-572-2206   Fax: (905)-572-2307

Denmark
Danmarks Laegelige Netvaerk, Trunnervangen 4a, 2920 Charlottenlund, Denmark
    Tel: 3582 3335   Fax: 3964 0131

Germany
Bundesanstalt für Arbeitsmedizin, Noldenstrasse 40–42, D-01134 Berlin, Germany
    Tel: 30 550 99 01/30 231 54 58   Fax: 30 231 54 31

Finland
Finnish Institute of Occupational Health

Ireland, Republic of

Health and Safety Authority, Temple Court, Hogan Place, Dublin 2   Tel: 6620400
    Fax: 6620417

Japan

National Institute of Health Sciences

Sweden

Arbetarskyddstyrelsen (ASS) , S-171 84 Solna, Sweden   Tel: 46 8 730 90 00
    Fax: 46 8 730 19 67

UK

Health and Safety Executive Books, PO Box 1999, Sudbury, Suffolk CO10 6FS, UK
    Tel: 01787 881165   Fax: 01787 313995
Health and Safety Executive Information Centre, Broad Lane, Sheffield, South
    Yorkshire S3 7HQ, UK   Tel: 0114 2892345   Fax: 0114 2892333
Health and Safety Executive, St Hugh's House, Stanley Precinct, Bootle, Merseyside
    L20 3QY, UK   Tel: 0151-951 4381 Personal callers only
Health and Safety Executive, Baynards House, 1 Chepstow Place, Westbourne
    Grove, London W2 4TF, UK   Tel: 0171-221 0870   Fax: 0171-221 0870

USA

National Institute for Occupational Safety and Health (NIOSH) , 4676 Columbia
    Parkway, Mail Stop 13, Cincinnati, Ohio, USA   Tel: 45226-1998   Fax:
    (513)-533-8573
Occupational Safety and Health Administration (OSHA), Salt Lake City, Utah, USA
    Fax: (202)-219-9266   http://www.osha..gov/safelinks.html   10 Regional
    Offices

For information relating to other countries consult:

The International Occupational Safety and Health Information Centre (CIS)
International Labour Office
CH.1211   Geneva 22
Switzerland
Tel: +41 22 799 6740
Fax: +41 22 798 6253
Telex: 415 647 ILO CH
E mail: 10043.2440@compuserve.com
Internet web site: http://turva.me.tut.fi/cis/home.html

The ILO produce the International Directory of Occupational Safety and Health
Institutions prepared from the CIS database giving up-to-date information on
responsibilities and practices of more than 300 key organizations operating in 93
countries.

Glossary

Absolute plating efficiency
The percentage of cells that give rise to colonies. See relative plating efficiency

Adherent (anchorage dependent)
Cells or cultures which will survive and maintain function only when attached to an inert substrate, e.g. plastic, glass, etc. The use of this term does not imply that the cells are normal nor that they are not neoplastically transformed.

Aneuploid
The condition which exists when the nucleus of a cell does not contain an exact multiple of the haploid number of chromosomes.

Aseptic
Minimizing risk of infection from bacteria, fungi or viruses.

Balanced salt solution
A solution containing a mixture of metabolically important inorganic salts in proportions designed to maintain the correct osmotic pressure.

Cell culture
Term used to denote the maintenance or cultivation of cells *in vitro* including the culture of single cells. In culture, the cells are no longer organized into tissues.

Cell fusion
The fusion of two or more dissimilar cells to form a synkaryon.

Cell line
A cell line emerges from a primary explant at the first successful subculture. Cell line implies that resulting cultures consist of generations of the cells originally present in the primary culture. See also Continuous cell line and Finite cell line.

Cell strain

Cells having specific properties or markers derived either from a primary culture or cell line are referred to as a cell strain; a characterized cell line derived by selection or cloning.

Cell transformation

A permanent alteration of the cell phenotype presumed to occur via an irreversible genetic change; may be spontaneous or induced by chemical of viral action, or by exposure to radiation. Transformation is often characterized (but not defined) by the emergence of an established cell line from a primary explant, an alteration in typical morphology, loss of contact inhibition, abnormal karyotype, changes in viral susceptibility, changes in antigenic properties, neoplastic properties, and the loss of anchorage dependency.

Chemically defined medium

A nutrient solution for culturing cells in which each component is of known chemical structure. Distinct from serum-free where other incompletely characterized constituents may be used to replace serum.

Clone

A population of genetically identical cells derived from a single parent cell by mitosis. A clone is not necessarily homogeneous and therefore the terms 'clone' or 'cloned' should not be used to indicate homogeneity in a cell population.

Cloning efficiency

The percentage of cells plated (seeded, inoculated) that form a clone. This term can only be used correctly where there is a certainty that the resulting colonies arose from single cells.

Confluent culture

Term applicable to adherent cells only where every cell except those at the very edge are in complete contact with other surrounding cells. By definition, it implies that no part of the substrate remains uncovered by cells.

Contact inhibition

Inhibition of cell membrane ruffling and cell motility when cells are in complete contact with adjacent cells, as in a confluent culture. Often precedes cessation of cell proliferation.

Continuous cell line

A population of cells which can be propagated for indefinite number of passages. Previously known as 'established' and often referred to as 'immortal'.

Crisis

A phase in the process of adaption by the cell to conditions *in vitro* where normal cell division pauses whilst genetic adaption occurs. The cell may or may not survive this phase which may last from several days to several weeks.

Density dependent inhibition of growth

Mitotic inhibition correlated with the increased cell density.

Differentiated

Cells that maintain, in culture, the specialized structures and functions typical of the cell type *in vivo*.

Diploid

A cell in which each chromosome is represented as a pair, and corresponding to the chromosome number and morphology of most somatic cells of the species from which the cells were derived.

Electroporation

Formation of transient pores in the cell membrane by means of an electrical current usually during transfection of DNA or other exogenous material into the cell from the surrounding medium.

Endothelium

An cell layer lining spaces within mesodermally derived tissues such as blood vessels, and derived from the mesoderm of the embryo.

Epithelioid

Implies cells derived from epithelium, but often used more loosely to describe any cells of a polygonal shape with clear sharp boundaries between cells. Pavement-like.

Epithelium

A lining of cells, as in the surface of the skin or lining of the gut, usually derived from the embryonic endoderm or ectoderm, but exceptionally derived from mesoderm as with kidney tubules and mesothelium lining body cavities. These cells frequently have a high nuclear to cytoplasmic ratio when compared to fibroblasts.

Eukaryotic cells

Cells in which the genetic material is contained within a distinct nucleus. This includes all organisms and cells except bacteria and blue-green algae (See Prokaryotic cells).

Euploid

A cell that contains exact multiples of haploid chromosomes.

Eutectic point
The minimum melting (or freezing) point of a mixture containing two or more compounds with different individual freezing points.

Explant
Tissue removed from an organism and placed *in vitro* in an artificial medium for growth. This constitutes a primary culture.

Feeder layer
The generation of a feeder layer involves a culture of normal adherent cell being allowed to grow to near confluency before being rendered incapable of further multiplication usually by irradiation. A fastidious cell type or cells at a low density can then be grown on this feeder layer which provides supplementation to the medium and may also condition the substrate. The presence of a feeder layer may also make it suitable for the attachment of certain other types of cells. For some applications, the feeder layer is separated from the cells by an inert perfusable matrix.

Fibroblastic
A proliferating precursor cell of the mature differentiated fibrocyte. These cells are usually spindle-shaped (bipolar) or stellate (multipolar) and are arranged in parallel arrays at confluence if contact inhibited. Cells are usually migratory with processes exceeding the nuclear diameter by threefold or more.

Finite cell line
A culture of cells which is capable of only a limited number of population doublings after which the culture ceases proliferation.

Generation time
The interval from one point in the cell division cycle to the same point in the cycle one division later. Distinct from doubling time or population doubling time.

Genotype
The total genetic composition of an organism: the combination of alleles.

Growth curve
A plot of cell number against time of a proliferating cell culture. Usually divided into *lag phase*, before growth is initiated; *log phase*, the period of exponential growth; and *plateau*, a stable cell count achieved when the culture stops growing at a high cell density.

Haploid
The chromosome number where each chromosome is represented once. In most higher animals, it is the number of chromosomes present in the gametes and half of

the number found in most somatic cells (symbol n). The basic number of a polyploid series (symbol x). Haploid in this meaning = monoploid.

Heterokaryon
Somatic cell hybrid containing separate nuclei from different species.

Heterologous population
A culture in which the cells do not all resemble each other. The culture contains morphologically different cells.

Heteroploid
A term used to describe a culture (not a cell) where the cells comprising the culture have chromosome numbers other than diploid. It does not necessarily apply that such cells are malignant or that they are able to grow indefinitely *in vitro*. In describing a heteroploid cell line, in addition to the karyotype of the stem line, the percentage of cells with such karyotype should be stated.

Homeostasis
Tendency towards a relatively constant state; equilibrium between an organism and its environment.

Homokaryon
Somatic cell hybrid containing separate nuclei from the same species.

Homologous population
A culture in which the cells are all identical (morphologically and genetically).

Hybridoma
The result of a fusion between two different cells. A hybridoma cell and its progeny contain some chromosomes from each fusion partner although some others are usually lost, e.g. (i) B-cell hybridoma: the result of a fusion between a normal plasma cell and a plasmacytoma cell. The resulting cell line continues to secrete the anti-body produced by the normal plasma cell which is monoclonal. (ii) T-cell hybridoma: a cell line obtained from a fusion between a T lymphoma cell and a normal T-lymphocyte.

In Vitro
Means literally 'in glass'. Usually refers to the cultivation of cells in a culture vessel or manipulation of cells external from the animal body. The opposite of *in vivo*.

In Vivo
Conditions found in the living animal.

Laminar flow
Layered monodirectional airflow. In a laminar airflow (LAF) cabinet air is forced through filters and specially designed ducting to produce an airflow with a minimal lateral disturbance. This has the effect of producing 'layers' of air where the adjacent layers do not mix except on the molecular scale. In the vertical LAF cabinet this airflow is in the form of a narrow 'curtain' of air passing down the inner front surface of the cabinet: in a horizontal LAF cabinet the air forms a 'block' the size of the front opening blown towards the operator.

Monoclonal antibody
Antibody produced by cloned cells derived from a single lymphocyte, usually a hybrid cell.

Monolayer
A single layer of anchorage-dependent cells growing on a substrate.

Mutant
A cell that has varied phenotypically from the original population owing to a modified or added gene.

Myeloma
A tumour derived from plasma cells (synonymous with plasmacytoma).

Passage
The process of transferring or transplanting a cell population from one culture vessel to another. This usually involves 'splitting' or dividing the cell culture by dilution. For cells which are not actively growing, subculturing may require reducing the cell to surface (medium) ratio. Synonymous with the term 'Subculture'.

Passage number
The number of times a population has been subcultured or passaged since it was established as a primary culture.

Phenotype
The observable characteristics of an organism as determined by the interaction of its genotype and its environment.

pK_a
The measure of strength of an acid defined as the negative logarithm of the acid ionization constant (K_a).

Plasma cell
A terminally differentiated B-lymphocyte. The major immunoglobulin-producing cell type.

Plasmacytoma
A tumour derived from plasma cells (synonymous with myeloma).

Plating efficiency
See Absolute plating efficiency and Relative plating efficiency.

Ploidy
The number of chromosome sets.

Polyclonal antibodies
Antibodies produced by a number of different lymphocytes as part of the normal immune response.

Population density
The number of cells per unit area or per unit volume in a culture vessel. Also called cell density.

Population doubling time
The interval in which a culture increases its total number of cells by a factor of two. This term is not synonymous with cell generation time.

Primary culture
A culture initiated from an explant of cells, tissues or organs taken from an organism. A culture is a primary culture until it is subcultured for the first time, at which point it becomes a secondary culture and is often referred to as a cell line.

Prokaryote
Cells in which the genetic material is not enclosed in a defined nucleus. Refers only to bacteria and blue-green algae.

Relative plating efficiency
The percentage of cells that give rise to colonies when compared to a control culture in which the plating efficiency is arbitrarily established at 100%. See Absolute plating efficiency

Seeding (attachment) efficiency
The percentage of inoculated cells which attach to the surface of the culture vessel within a specified time.

Subculture
Harvesting cells from one culture and using them to initiate a new culture. The term is synonymous with passage.

Subculture interval
The time between sequential subculturings of a cell line. Synonymous with passage interval.

Suspension culture
A culture in which cells have been adapted to grow and divide without being attached to a substrate.

Syncitia
Multinucleated cells.

Tissue culture
The study of cells, tissues and organs maintained *in vitro* for more than 24 hours. Three main branches of tissue culture are recognized:
Cell Culture, the growth *in vitro* of cells, including the culture of single cells, no longer organized into tissues.
Tissue Culture, the maintenance of tissue fragments *in vitro;* original or characteristic architecture is not necessarily maintained.
Organ Culture, the maintenance or growth of tissues, organ primordia, or the whole or parts of an organ *in vitro*; designed to allow differentiation and maintenance of the original architecture and/or function.

Transfection
The transfer, by artificial means, of genetic material from one cell to another. Implies transfer of less than the whole nucleus of the donor cell and is usually achieved by using isolated chromosomes, DNA or cloned genes.

Transformation
A permanent alteration of the cell phenotype presumed to occur by an irreversible genetic change. May be spontaneous, as in the development of rapidly growing continuous cell lines, or induced by chemical or viral action. Usually produces cell lines which have an increased growth rate, an infinite lifespan, a higher plating efficiency and are often tumourigenic.

Zwitterionic buffer
From the German word meaning hybrid. They contain both positive and negative ionizable groups: secondary and tertiary amino groups provide the positive charges whilst sulphonic and carboxylic groups carry the negative charges.

Index